Public Spaces for Water

Sustainable Cities Research Series

ISSN: 2472–2502 (Print)
ISSN: 2472–2510 (Online)

Book Series Editor:

Bruno Peuportier

Centre for Energy Efficiency of Systems, MINES Paristech, Paris, France

Volume 3

Public Spaces for Water

A Design Notebook

Maria Matos Silva

*Universidade de Lisboa, Faculdade de Arquitetura,
CIAUD, Centro de Investigação em Arquitetura,
Urbanismo e Design, Lisboa, Portugal*

CRC Press
Taylor & Francis Group
Boca Raton London New York Leiden

CRC Press is an imprint of the
Taylor & Francis Group, an **informa** business

A BALKEMA BOOK

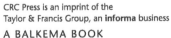 

CRC Press/Balkema is an imprint of the Taylor & Francis Group, an informa business

First issued in paperback 2021

© 2020 Taylor & Francis Group, London, UK Typeset by Apex CoVantage, LLC

Library of Congress Cataloging-in-Publication Data
Applied for

Published by: CRC Press/Balkema
            Schipholweg 107C, 2316 XC Leiden, The Netherlands
            e-mail: Pub.NL@taylorandfrancis.com
            www.crcpress.com – www.taylorandfrancis.com

ISBN-13: 978-0-367-03100-8 (hbk)
ISBN-13: 978-1-03-208263-9 (pbk)
ISBN-13: 978-0-429-02042-1 (eBook)
DOI: https://doi.org/10.1201/9780429020421

Contents

4   Discussion   123

Illustrations

Figures

Table

Acknowledgments

This work was supported by the individual doctoral grant provided by the Portuguese Foundation for Science and Technology (SFRH/BD / 76010 / 2011) as part of the POCH – Human Capital Operational Programme in conjunction with the European Social Fund and National Funds of the Ministry of Education and Science, and by Centro de Investigação em Arquitetura Urbanismo e Design (CIAUD) from the Faculty of Architecture, University of Lisbon. The author further wishes to thank the Portuguese Association of Landscape Architects (APAP) for the endorsement of this publication.

Preface

The urban phenomenon of floods is recurrent and is expected to be aggravated in the near and distant future, not only in light of climate change projections but also if flood management approaches continue the path of "business as usual." Facing this problem, the research underlying this book proposes the design of public spaces as a key component in the adaptation to current and expected urban flood events.

Climate change adaptation endeavors have already entered the urban agenda and are influencing urban planning and public space design approaches. This emerging tendency is further prompting new flood management paradigms that acknowledge the practice of integrating ecosystems and the natural water cycle.

Public Spaces for Water: A Design Notebook develops from a solution-directed process, which is particularly attentive to design and envisions a direct application in contemporary practice. Overall, it aims to offer a wide range of systematized conceptual solutions, in order to promote and facilitate the initial stages of a public space project with flood adaptation capacities. These solutions, organized in categories and types of measures, are identified and characterized through simple technical notions, straightforward vocabulary and drawn schemes, offering a synthesized overview of ideas for those dealing with a public space design project with flood adaptation purposes. A significant sample of real-case situations is additionally considered, showing case examples applicable to different contexts and with different purposes, providing valuable information that may further assist the decision making throughout the design processes.

Entailing part of the contents of the author's doctoral dissertation presented in 2006, this book displays a different approach to tackle the urgent problem of urban flooding. Through a different perspective, one that highlights the importance of public space design in adaptation action, traditional flood risk management practices are confronted and improved. Conventional responses, practiced through singular and segregated disciplinary approaches, are therefore reassessed with the rich, wide-ranging and interdisciplinary benefits brought by public space design.

While science and global "top-down" approaches provide important knowledge, local "bottom-up" approaches provide critical wisdom. The quality of future cities can be influenced by the quality of adaptation measures applied in public spaces. As public spaces provide the opportunity to integrate and reveal the complex connections between natural, social and technical processes, designing "for water" is a present and urgent matter on which the future of our cities relies.

Foreword

The world has entered a new era known as the Anthropocene, in which humans are recognized as the primary force that is shaping, and changing, an increasingly urban world. In this "Century of the City" global urban population will reach 70% by mid-century. With this unprecedented urbanization and human influence come profound and novel challenges. Chief among these is adapting to climate change. In this book, Maria Matos Silva addresses these themes and challenges explicitly through the "lenses" of water, climate adaptation, and design.

Water is essential for all forms of life and is the universal solvent. More than any other resource, water integrates humans with their environment. As it moves across and through cities, water responds in real-time to the built environment and transports virtually everything it contacts to "downstream" locations. In a context of climate change, urban water exists across a continuum from scarcity to flooding. And the quality of urban water continues to suffer from contamination and pollution. Maria Matos Silva summarizes these challenges with a timely and existential proposition "designing for water is a present matter in which the future of our cities relies!".

This design notebook takes on the profound challenges of urban water and hydrology from several perspectives: theoretical, professional and global case-studies. The theory of designing public spaces for water benefits from Matos Silva's recent dissertation research. She supports the book with research on the Anthropocene, climate change, uncertainty and the imperative for an adaptive response. Appropriately, she integrates the social dimension into her theory and to provide a framework for "bottom-up" public engagement and interaction with public spaces for water. The core of the book is a comprehensive library of flood adaptation measures applicable in the design of public spaces. Here designers will find measures for virtually any context or application. The measures are illustrated with clear and informative diagrams using a familiar professional graphic style. The last section of the book covers a highly international set of case studies where the measures of the notebook have been effectively implemented. The case studies are highly international and diverse in context and project goals/purposes.

Readers of the notebook will learn how to design public spaces for water that adapt to new climate realities in cities across the world. In this "Century of the City" public spaces are expected to provide more functions – particularly with respect to climate change. Urban plazas can serve as reservoirs for runoff, mitigating occasional flooding while also supporting the culture and social life of the city. The many new ideas presented in this notebook are guided by contemporary design principles including: interdisciplinarity, public engagement and shared values, and explain how to manage and diversify risk and "learn-by-doing" through performance monitoring.

Managing water in cities is one of the great challenges of our time – with no exceptions. Every city has a unique version of the urban water management challenge. In this "design notebook" Maria Matos Silva gives urban designers, landscape architects, policy makers and civic leaders a valuable body of knowledge and examples to address this challenge – deeply rooted in contemporary sustainability and resilience theory and based on emerging international best practice.

Jack Ahern, Ph.D., Professor of Landscape Architecture,
University of Massachusetts, USA

Introduction

Floods are a recurring phenomenon that disrupt our cities on a regular basis; they can cause environmental damage, displace populations, and even cause the loss of lives. Floods can also seriously compromise economic activities and economic development. Changing this rising tendency is a massive challenge, particularly when considering the continued research that has been consolidating knowledge on climate change projections, pointing toward an even greater number of flood events per year. Science periodically evidences that flood events are likely to increase not only due to projected sea level rise but also due to an expected increase in intensity and frequency of extreme storm events.

Existing flood management infrastructure has been designed in accordance with statistical recurrence criteria, originating from available historic meteorological data, yet the same method may not be able to cope with future flood occurrences, which are expected to be unprecedented. Although we have always lived and will continue to live with a changing climate, it is very likely that our cities are not prepared to face the projected exceptional changes, as our infrastructure relies on obsolete and past determinations. Even assuming that most of the existing infrastructure is inadequate, altering flood management systems from one moment to another in order to follow projected climate change scenarios would involve unreasonable investments, which are very likely unaffordable for municipalities in the near future.

Fortunately, progressive adaptation to projected flood events is gaining relevance as one of the biggest challenges of the present century. Notwithstanding the unquenchable controversy that forever follows climate change science, the concern about possible devastating impacts – and the consequent need to act upon them – is high on the international political agenda. Global Risks Report of 2016, presented at the World Economic Forum, even presented the "failure of climate-change mitigation and adaptation" as the major "global risk of highest concern" for the next 10 years, considering it to have a greater damage potential than "weapons of mass destruction," placed as the second most important risk before "water crisis" in third.

The proposed work derives precisely in the face of this contemporary concern, which is acknowledged by more people every day as a global emergency, and specifically upon adaptation action toward one of the most impactful threats, that of urban floods.

Recognizing that present flood management infrastructures are not enough to respond to the scale and speed of estimated flooding impacts, flood management practice must urgently reassess and improve established methods. The challenge put forward by climate change, which is pushing infrastructure to its limits, is forcing established

disciplines and professions to redefine their competences and pursue structured interdisciplinary collaborations, moving beyond sectorial and self-absorbed paradigms. It is in this
line of reasoning that the design of public spaces is faced with a new and enhanced role
as a mediator for flood adaptation that is able to integrate infrastructure and communities
together in the management of floodwater as an ultimate resource for urban resilience.

The present relationship between urban planning and flood management infrastructure
generally consists of occasional actions driven by political priorities, with potential solutions carried out only when complications are already in a severe state. This type of
approach is unsuitable considering future projected scenarios. Estimated projections are
indicative rather than definite, yet they are reliable as such. The fundamental ideas that
infrastructure should be integrated into urban planning and that urbanism should better consider estimated projections as drivers of change are not new. What is here proposed is that
by specifically introducing public space design in all planning stages of flood management
infrastructures, significant and successful progress in the agenda of urban flood adaptation
action are possible.

Public spaces are here understood as multifunctional spaces, with a central social, political and cultural significance. They are distinguished by their long-lasting permanence as a
structuring urban space, and they have an interdisciplinary nature. Implicitly, they are considered as a common entity of shared concerns, which may also accommodate civic purposes. Moreover, public spaces are among the urban areas most vulnerable to climatic
hazards, particularly flooding events; simultaneously, they entail specific characteristics
that are particularly relevant for adaptation efforts. Indeed, not only are hazards are
more acutely felt at the local level, but it is also local communities that have the most
know-how and experience to promptly deal with existing vulnerabilities. By integrating
flood management infrastructures as part of the design of public spaces, what was previously strictly seen through the lenses of single and isolated disciplines is forced to
embrace the core of an interdisciplinary design process: a process where local climate
changes can be made visible and consequently meaningful for citizens and their livelihoods; a process that furthermore provides opportunities for the experiential learning, monitoring and ongoing management, so fundamental to any adaptation endeavor.

One may identify a wide range of current and past situations where a public space is
combined with one or more flood management measures. One may furthermore witness
more recent examples where public spaces and flood risk management measures are explicitly integrated with climate change adaptation purposes. These examples encourage and
promote new ways of thinking, new solutions and, eventually, new paradigms. However,
there is still a considerable amount of unconnected information between theoretical findings and a professional exercise of comprehensive application. The question remains on
"how" to expand the matter of urban flood adaptation into generalized practice. This
motion steered this book.

For those involved in the design of public spaces that incorporate flood adaptation
efforts, analyzing the existing knowledge is crucial. And indeed, flood adaptation measures
applied in the design of public spaces have been sporadically discussed by a range of
diverse literary references. However, analyzing the existing knowledge must be done in
a systematic way; otherwise, it may present unsatisfactory results when considering the
time frame of one design proposal. This book aims to facilitate the initial phases of a
design process by exploring and systematizing a range of flood adaptation measures applicable to the design of public spaces, alongside real case examples. Different categories and

subsequent adaptation types of measures are specifically identified and characterized through simple technical notions, straightforward vocabulary and drawn schemes. Overall, the book envisions to offer a different approach to tackle present and future impacts of urban flooding, one that highlights the importance of public space design in adaptation action.

Chapter 1

Public spaces for water

Introduction

Timeframes of significant climate variability have occurred since the earth's origin and through its habited ages. Humans, often characterized as the most adaptable of animal species, have always been able to successfully adapt to altered climates. Moreover, throughout most of the human history, we had a residual interference in the earth's natural systems, at least until the Industrial Revolution.

There have been several authors who, since the modern environmental conservation movement, have argued that the period commonly identified as the Industrial Era may be the cause of radical climatic changes. Among them are George Perkins Marsh, who wrote *The Earth as Modified by Human Action* (1882), a pioneering book on conservation science; Nathaniel Shaler's book *Man and the Earth* (1905); and Rachel Carson's *Silent Spring* (1962), among others (Hebbert and Jankovic, 2013).

In 1975, Wally Broecker specifically engaged with the matter of anthropogenic global warming. Through his paper "Climatic Change: Are We on the Brink of a Pronounced Global Warning?" Broecker was one of the first to argue that by the first decade of the 21st century, global temperatures would be warmer than any in the antecedent millennium (1975).

Later by the 1980s, sociologists Ulrich Beck and Anthony Giddens termed the concept of "risk society" when reflecting upon modernity, and in particular, the growing environmental concern. For Giddens, a risk society is "a society increasingly preoccupied with the future (and also with safety), which generates the notion of risk" (Giddens, 1999a, p. 3). At the same time, Beck describes risk as "a systematic way of dealing with hazards and insecurities induced and introduced by modernisation itself" (Beck, 1992, p. 21).

The severe ecological disruptions that have resulted from industrial society, such as the climatic changes emphasized by the authors mentioned earlier, served as a key pillar for this analysis of the modern period. For Beck, environmental risks have become recurrent rather than exceptional. The author further argues that this occurrence is a result of "manufactured constraints," or, in other words, the result of pressures that are significantly determined by human actions (Beck, 1992, p. 175). In order to clarify the difference between external or "natural" risks and "manufactured risks," Giddens elucidated that, "At a certain point, however – very recently in historical terms – we started worrying less about what nature can do to us, and more about what we have done to nature" (Giddens, 1999b, p. 3).

More recently, one of the most important reflections on this matter may be the suggestion made by Nobel Laureate Paul Crutzen and Eugene Stoermer in 2000, that we have entered a new geological era after the Holocene. They coined the era as the Anthropocene and defined it as an unprecedented age in which humans are not just mere spectators but the primary forces shaping the world (Crutzen and Stoermer, 2000).

In 2002, Crutzen developed from his original article with a commentary in the journal *Nature*, the "Geology of Mankind," stating that "The Anthropocene could be said to have started in the late Eighteenth century when analysis of air trapped in polar ice showed the beginning of growing global concentrations of carbon dioxide and methane" (Crutzen, 2002, p. 23).

Among other consequences, Crutzen and Stoermer argue that humankind has exhausted 40% of the known fossil fuels in only a few generations; nearly 50% of the land surface has been transformed by direct human action, from the dams holding sediment by the gigaton to the forests' devastation; more nitrogen is now fixed synthetically for fertilizers than is fixed naturally in all terrestrial ecosystems; there are now less than 50% of mangroves protecting coastal wetlands; fisheries remove more than 25% of the primary production of the oceans; and more than half of all accessible freshwater is used for human purposes (Crutzen and Stoermer, 2000, p. 17).

The idea that we have transitioned to a new geological era where humanity is the main influencer has a far greater reach than a simple change of name. It implies a new perspective on how to manage the relationship between people and earth. It further implies that a new way of thinking and acting is urgent. This existential analysis we are forced into goes in line with the concept of "modern reflexivity," also widely argued by Beck and Giddens. For Beck, the key feature of "reflexivity" is the process of society examining itself and acting accordingly: "what was made by people can also be changed by people" (Beck, 1992, p. 157).

In 1988, the World Meteorological Organization (WMO) and the United Nations Environment Programme (UNEP) established the Intergovernmental Panel on Climate Change (IPCC). Since then, this entity has been producing, at regular intervals, assessment reports (AR) on the state of knowledge on climate change. During an assessment experience of a quarter of a century, there were substantial progressions among the published results. Experts contributing and assessing these reports also substantially increased in number. While the FAR had the contribution of 97 authors, the AR4 received contributions from over 3500 experts from more than 130 countries (IPCC, 2015).

In 2013 there was a 97% consensus rate among climate experts in regard to anthropogenic global warming (Cook *et al.*, 2013, p. 6). The remaining 3% not only question the consensus of climate scientists but also the consensus of evidence, which is mostly argued by the uncertainty associated with climate research. Sources of uncertainty have been identified as arising from measurement errors; aggregation errors; natural climate variability; future emissions of greenhouse gases (GHG); limited climate models; complexity in interaction of climatic and non-climatic factors; and future changes in socio-economic, demographic, and technological factors, as well as in societal preferences and political priorities (EEA, 2012a, p. 42). Other concerns include the vulnerability of systems and regions, the conditions that influence vulnerability and particular attributes of adaptation, such as costs of implementation and maintenance, effectiveness, and significance (Burton *et al.*, 2001). For example, while the potential contribution of ice sheets to sea level rise (SLR) is very large, there are still many incomprehensible processes concerning their dynamics. As stressed by Michael Oppenheimer regarding this matter, "uncertainty is still large, and is unlikely to ever be reduced," and "it is also likely that, despite the enormous progress, the phenomena of the 21st century will anticipate their proved predictions" (2010,

p. 12) – arguments that not only question the existence of forthcoming perfect models but that also highlight our present unpreparedness in regard to projected impending weather events.

While in theory some uncertainties may be reduced by further research, others simply cannot, as they are related to the future. Indeed, there is no way around uncertainty in climate change studies. There is no way around the fact that no research will never be exact about the future. Scientists will only be capable of proposing ranges of partial or imperfect information that indicate approximate tendencies. Regardless, "climate skeptics" will always haunt climate change action.

Several authors and near-universal agreement[1] have strongly argued for the need for countries to invest on increasing their capacity to cope with uncertainty rather than to increase risk through the use of ambiguous impact studies or no action (Intergovernmental Panel on Climate Change (IPCC), 2012, p. 351). As stated by the British philosopher and logician Carveth Read, "It is better to be vaguely right than exactly wrong" (Read, 2012 [1898], p. 351).

Acknowledging that uncertainty about future weather events is a sure certainty, the ultimate question is how we can prepare and manage for our future. Particularly in regard to the transposition of this problem upon urbanism, and in light of the concept of Ulrich Beck's "risk society," François Ascher was one of the first to argue that urban planning had to deal with uncertain risks and global impacts (Ascher, 2010[2001]). As stated by Richard Marshall,

> Our cities have changed faster than we have been able to adjust our thinking. (…) Our problem is not one of memory; it is one of adjusting our ideas of what is an appropriate urban form to be in line with the current reality of our culture and society. What is needed in urban design today, above all else, is a re-calibration of our ideas to the currency of our time.
>
> (Marshall, 2001, p. 3)

For Lister, "if uncertainty and regular change are inevitable, then we must learn to be flexible and adaptable in the face of changes" (Lister, 2005, p. 21). Following this line of reasoning, Jack Ahern further argued that uncertainty must be reconceived as an opportunity to "learn by doing" (Ahern, 2006, p. 129). In accordance with these arguments, designers and planners, which operate in the "real world," cannot be tied up until there are no more climate skeptics or the hopelessness of no more uncertainties.

In line with the research project Urbanised Estuaries and Deltas (Costa et al., 2013), however uncertain, climate change projections provide a sufficiently stable range of possible futures that serve to test available options. The methodological difference upon the planning process relies mainly on the shift from the search for the one optimal solution into the pursuit of various adequate alternatives.

I   Climate change adaptation through local, "bottom-up" initiatives

Climate change has mostly been evaluated through global models, more specifically, through general circulation models (GCM), in order to anticipate climate change scenarios. However, many authors strongly support scientific evidence claiming that "changes in climate are happening at multiple scales from global to regional to local and that there are independent anthropogenic drivers of change at each scale" (IPCC, 2007; Oke, 1997; Stedman, 2004 *in* Ruddell *et al.* [2012, p. 584]). As a result, regionalized models (RCMs, regional climatic models) started being used. These models derive from the downscaling of the GCMs and cover a limited area of interest, such as Europe or an individual country. As a consequence of being based on an incomplete model *per se*, these regionalized models reinforce eventual errors and insufficient data. As stated by Hebbert and Webb, "[climatic] effects cannot be downscaled from a regional weather model, they are complex and require local observation and understanding" (2007, p. 125). Global as well as regional models, particularly when considering the necessary combination of overwhelming information about all the natural and changing processes, are therefore further distanced approximations of reality.

In addition, although it is commonly recognized that great driving forces function at a global scale, such as greenhouse gas rates or financial dynamics, it is also widely acknowledged that various local phenomena influence global climate (Wilbanks and Kates, 1999, p. 602), from micro-environmental processes to demographic variations or resource-use undertakings, such as deforestation or coral mining. The urban heat island (UHI) effect, in particular, as a clear indication of significant acceleration of temperature changes in most existing cities worldwide, imposes direct repercussions upon global climate.

Other authors have further argued that, although the frequently mentioned greenhouse gas emissions unequivocally contribute to global warming, the witnessed disturbance of the small water cycle is a bigger catalyst on future climate extremes (Kravčík *et al.*, 2007, p. 7). While most investigations analyze the impacts that climate change will have on the water cycle, Kravčík *et al.* question the reverse influence that an unbalanced water cycle may have on the exacerbation of climatic change. In light of the research presented by these authors, saturating the small water cycle through the conservation of rainwater on land would be a revolutionary solution to the given problems of anthropogenic climate change.

For a long time, carbon mitigation and/or adaptation to global warming have formed part of several international agendas, with monthly initiatives being disseminated throughout the global scientific community. Although these global endeavors are imperative, it has been argued that it is frequently "focussed [on] the exposure of cities to hazards that have a huge impact but low frequency. It has little to say about the high-frequency and micro-scale climatic phenomena created within the anthropogenic environment of the city" (Hebbert and Webb, 2007, p. 126). Contrastingly, this tendency has been counterbalanced by various local undertakings such as the water saving projects in Zaragoza, the establishment of community-based early warnings against flash floods in northern Bangladesh or the neighborhood's action against the impact of urban heat islands in Portland, Oregon (Ebi, 2008 in Intergovernmental Panel on Climate Change (IPCC), 2012, p. 321). These initiatives essentially confront the systematic assumption of realism in science (as highlighted by Beck, 1992, p. 5), deciding not to rely solely on justifications from global projections in order for adaptation to be advanced in cities. Rather than being restrained or expectant of downscaled or locally applied models, a wide range of cities have recognized the need to take action now in order to prepare for the future (Carmin *et al.*, 2012).

In this line of reasoning, Jaap Kwadijk and others (2010) have identified two main approaches on climate adaptation policy: (1) a predictive top-down approach and (2) a more from the bottom-up resilience approach. While the first is essentially guided by global models, the second is relatively independent of "justifications from atmospheric science" (Ruddell *et al.*, 2012, p. 601) and its associated uncertainties. Moreover, instead of reducing impacts, the latter rather focuses on reducing vulnerability by improving the resiliency of a system exposed to particular climate change risks (Te Linde, 2011 in Veelen, 2013).

Local scales are particularly sensitive to every climatic change, be it sporadic or ongoing. According to the IPCC, there is "high agreement" and "robust evidence" that "disasters are most acutely experienced at the local level" (Intergovernmental Panel on Climate Change (IPCC), 2012, p. 293). When analyzing the risks of extreme events, the IPCC further highlights that while most events will not become severe enough to cause a disaster of national of international magnitude "they will create ongoing problems for local disaster risk management" (Intergovernmental Panel on Climate Change (IPCC), 2012, p. 297). The degree of the impacts is also strongly linked with the existing social and physical local vulnerabilities "including the quality of buildings, the availability of infrastructure, urban forms and topographies, land uses around the urban centre, local institutional capacities" (Bicknell *et al.*, 2009, p. 362).

Furthermore, as locals are the first to experience and respond to hazards, they retain local and traditional knowledge that is not only aware of these hazards and existing vulnerabilities but also knows how to cope and work with them (Intergovernmental Panel on Climate Change (IPCC), 2012, p. 298). In accordance, Ruddell *et al.* states that "It is critical to ground support for climate adaptation and mitigation initiatives within local contexts of shared experiences" (2012, p. 601). In other words, local know-how should always be considered as added value for adaptation action, particularly when considering a known or often repeated hazard. Societal processes are thus critical for the success adaptation action. Ultimately, and regardless of national and reginal efforts, without a local approach that integrates the human scale, adaptation endeavors will fail their purpose.

On the other hand, projected scenarios and the increased record of more frequent extreme events (Coumou and Rahmstorf, 2012) will likely lead to unprecedented situations for which localities have no previous experiences. As corroborated by the IPCC, "extreme weather and climatic events will vary from place to place and not all places have the same experience with that particular initiating event" (Intergovernmental Panel on Climate Change (IPCC), 2012, p. 297). As such, local action should not be dissociated from a global and more encompassing scale in which scientific knowledge of future climate projections is crucial. For example, bearing in mind that local initiatives rely more on weather, ecology, and social media, events such as cold summers or heavy rains may be wrongly interpreted as evidence that there will be no global warming or long term droughts (Ruddell *et al.*, 2012).

In accordance, considering the risk of not contributing effectively to the achievement of community expectations and the safeguarding of public interests and collective resources, not only local but also global and regional strategic views must be taken into consideration. In other words, both global "top-down" and local "bottom-up" strategies must be equally significant drivers for successful adaptation action. In the same line of reasoning, in the final conference of the R&D project "Urban Deltas" (Costa *et al.*, 2013), Han Meyer has raised the question of how to take advantage of the arising numerous local initiatives. In his view, local actions that are not structured in one global and general strategy get lost in the overall scale and will ultimately lose their significance and value. Indeed, while acknowledging the importance of local-scale responses, these should always be accompanied by a global strategy in order to fulfill their objectives.

International examples, such as the Netherlands, the United States, and England, and more specifically the cities of Rotterdam, New York, and London, have taught us the common practice of approaching flood adaptation locally, comprehensibly valuing and favoring community engagement and involvement. In light of these examples, it was possible to verify how agile municipalities, which often have close relationships with their citizens, enterprises and institutions, are quicker and more effective in the implementation and monitoring of local adaptation solutions – recalling Borja, "it is not possible to decouple urban claims from the strength and innovation of local and proximity governance" (Borja, 2003, p. 31, author's translation). On the other hand, for Bicknell, to ensure adaptation to extreme weather events is a characteristic of well-governed cities (2009). Hebbert and Jankovic further stated that "Cities which understand and manage their local climate have a head start in responding to global climate change" (2013, p. 1345), a "municipalism in action" that strengthens the social, physical and economic backbones of the city (RCI, 2009) but that is also inseparable from local competences and political autonomy.

In addition, as has been argued by many authors, there is a strong relation between the quality of a city and the quality of its public space. This relation can be acknowledged on both sides of the spectrum, i.e., that the quality of a city can be measured by the quality of its public space and that the quality of public space largely influences the quality of cities. More specifically, for John Ruskin, "The measure of any great civilisation is in its cities, and a measure of a city's greatness is to be found in the quality of its public spaces, its parks and its squares" (Cowan, 2005, p. 314), and for Jordi Borja, "The assessment of urbanism is the public space" (Borja, 2003, p. 176, author's translation). On the other hand, Jordi Borja also emphasized that "Public space defines the quality of the city, because it reveals people's quality of life and the quality of citizenship for its inhabitants" (Borja, 2003, p. 135, author's translation). Furthermore, Brandão et al. argued that "Quality public spaces can help cities to create and maintain sites of strong centrality, environmental quality, economic competitiveness and sense of citizenship" (Brandão et al., 2002, p. 17, author's translation). Other widely recognized authors have considered that the city "is" the public space itself (Lynch, 1996[1960]; Jacobs, 1992[1961]; Portas, 2011[1968]) and "the city is the public space, place of social cohesion and exchanges" (Borja, 2003, p. 119, author's translation).

What came to be known as the Barcelona model is a widely recognized example of a successful urban regeneration process that is essentially focused on the improvement of local-scale public spaces. As evidenced by its greatest mentor, Oriol Bohigas, the main focus was on the improvement of the quality of public spaces, either as a result of public or private initiatives. Following the general approach of "cleaning the centre and monumentalize the periphery" (Bohigas, 1986, p. 20), any intervention was primarily based on the (re)qualification of public spaces or on the building of new ones. From the PERI projects (Special Plans for Interior Reforms) (1980–1986) to the Olympic project (1982), the Barcelona model gave rise to a new form of planning based on the importance and dignity of public space. It was the emergence of Barcelona after Franco. In the first democratic urbanism projects, public space started to be something belonging to all citizens, which everyone needs and that everyone must have access to.

Eventually, one may identify a current "flip side" to the successes achieved in the 1980s and 1990s, namely as a result of the negative consequences of an over-growth of the tourism sector and overall "globalization" (Remesar, 2005). Yet what made the case of Barcelona a model was the generalized perception of the ideal to base urban regeneration process on interventions made with public money in order to create new public spaces for its citizens. As mentioned by Balibrea, from the moment that public intervention obeys the institutional relations logic of the local situation of global markets, and not of those institutions with the needs and desires of the local population, the meaning of public space undermines itself, and its continued evocation can turn out to be only a rhetorical speculation (Balibrea, 2003).

It is interesting to note how nowadays one can still verify reminiscences of the aforementioned early democratic urbanism ideals, namely in the management of the cities' floods. In similarity to most developed cities, what were formerly open and exposed water courses – which crossed Barcelona's plain, parallel to each other, into the sea – are now underground interconnected channels. As the urbanized impermeable areas increased, the drainage system progressively became insufficient to manage the heavy rains experienced. Consequently, Barcelona increasingly suffered from recurrent flooding episodes.

In addition, besides the torrential precipitation common to Barcelona's Mediterranean geographic condition, the topographical situation of the city is also a significant and inevitable contributing factor for the occurrence of periodic flash floods, namely the steep slopes of the Collserola mountain and the cities' extended plain surface in between the mountain and the sea. While the steep slope doesn't allow infiltration, the extended plain does not favor gravity drainage. Consequently, there is a rapid concentration of rainwater into the lower lands and a subsequent difficulty to discharge stormflows to the receiving waterbody. Recall how in 1860, Cerdá's plan took this matter into consideration and "adapted" the "Ensanche" itself to the northwest-southeast/northeast-southwest orientation in order to facilitate drainage runoff.

Nowadays, most of Barcelona operates in a combined sewer system (CSS). In dry weather and during light to moderate rainfall, the system conveys wastewater and stormwater flows to the wastewater treatment facility. However, during periods of heavy rainfall, the capacity of the CSS and its treatment plant can be exceeded, and combined sewerage is diverted onto a receiving waterbody. This phenomenon is called a combined sewage overflow (CSO), which not only contributes to the occurrence of floods but also directly impacts the receiving waterbody with untreated polluted waters.

In order to tackle this matter, several retention tanks were proposed in the 1997 "Plan Especial de Alcantarillado de Barcelona" (PECLAB) and were later constructed by Clavegueram de Barcelona, S.A. (CLABSA). These great regulation tanks or underground reservoirs had the primary function of retaining vast quantities of stormwater during heavy rains. Once large amounts of stormwater are retained, they can be subsequently evacuated little by little, and the system is ready to receive them. Through such a process, untreated outflows into the beaches and rivers is also avoided.

In 2011, Barcelona had 11 reservoirs that together could store around 415,000 cubic meters of stormwater (CLABSA), an investment that greatly reduced the flood episodes within the city. Yet what makes Barcelona's reservoirs particularly different from the others implemented in many European cities, such as Berlin (Germany), Bolton (United Kingdom), or Bordeaux (France) (ChiRoN et al., 2006), is their integration with cities' public spaces.

Barcelona is often highlighted as being the European city whose drainage strategy was invested in the most, through construction of these underground reservoirs. However, the city should also be recognized for its capacity to integrate such great infrastructures in compact urbanized areas as well for its capacity to combine synergies in order to offer more public spaces for its citizens.

Figure 1.1 Some of Barcelona's public spaces over underground stormwater retention reservoirs. From left to right and top-down: Doctor Dolsa square; Escola Industrial sports field, Nou Barris park and Joan Miró garden.

Source: Author's personal archive, 2011.

Most of the constructed underground reservoirs in Barcelona encompass squares, sports fields, parking lots or gardens on their surface (Figure 1.1). Through the construction of these reservoirs, Barcelona has shown that infrastructures may aspire to more than only serving the objective they were built to accomplish. In contrast with other cities that look into infrastructure as an isolated and untouchable urban element, the project and implementation of such reservoirs interconnected several disciplines, from urban planning to drainage engineering and landscape design, a transversal and interdisciplinary approach that explored the benefits and opportunities of common understandings. In 2003, Jordi Borja had already emphasized that

> infrastructures have been considered as an inevitable aggression to the citizen's public space or have not been treated for other uses besides the specifics of their roles: network services (energy, water, telephone, etc.), infrastructures and collective transport systems (from train stations to bus stops) [but that] one needs to see these elements as opportunities and not obstacles in the development of the city and its quality of life.

(Borja, 2003, p. 137)

It thus appears that the importance given to public space, as a distinctive feature of the Barcelona model, is latent in contemporary sectorial strategies and not only in the past initial direct intent of urban regeneration. It may be further argued that the Barcelona model therefore remains a pioneer urban approach, namely on its combined integration between flood adaptation measures and public space design. Although Barcelona's reservoirs were not conceived with specific adaptation purposes, they are undoubtedly connected to a particular kind of public space that shares the additional function to tackle floods. Through their public spaces, people are connected with the dynamics of a particular system or infrastructure. More specifically, physical elements such as ventilation chimneys or placards evidence the presence and performance of the belowground infrastructure. This means of connection between people and the flood management infrastructure may lead to an improved common understanding and communal engagement.

Lastly, recalling upon the previously exposed advantages of locally driven adaptation action, it is argued that just as specific public space interventions can rise the standard for good quality cities, so too will local adaptation measures applied in public spaces be particularly relevant in the quality of our future cities. The remaining part of this chapter will continue to reason upon this subject as also presented in (Matos Silva and Costa, 2018).

2   The key role of public space in adaptation endeavors

> To live together in the world means essentially that a world of things is between those who have it in common as a table is located between those who sit around it; the world, like every in-between, relates and separates men at the same time./The public realm, as the common world, gathers us together and yet prevents our falling over each other, so to speak.
>
> Arendt, H., 1998[1958]. *The Human Condition*.
> 2nd Edition ed. Chicago University of Chicago Press, p. 52

The idea of public space may be apprehended by a set of two meanings: (1) a conceptual meaning commonly used in political and social science, in which public space, as characterized by Innerarity, "brings together all the processes that configure the opinion and collective will" (2006, p. 10, author's translation) and (2) a physical meaning, commonly used in urban planning and design, where the previous actions are developed (Borja, 2003, p. 22; Cowan, 2005).

In old Greece, "public space" had specific boundaries. It was called "Ágora," whose literal meaning is a "gathering place" for discussion. Innerarity has argued that this traditional meaning of public space, in which public matters are expressed and represented in a bounded physical area, has disappeared. However, the author additionally claims the need of spaces "for" the public (Innerarity, 2006, p. 135) – more specifically, spaces that enable and promote community life, such as streets, sidewalks, squares, coffee shops, parks or museums, and that potentially offer wide-ranging benefits such as place-making, sense of place or local identity. Likewise, other authors have additionally highlighted public space as multifunctional space, with a central social, political and cultural significance (Ricart and Remesar, 2013, p. 6). In regard to its physical characteristics, it has been particularly argued on the long-lasting permanence of public space as a structuring urban space (Martin, 2007; Portas, 2003) of interdisciplinary nature (Madanipour, 1997[1972]; Brandão, 2004). Overall, public space may be defined by Hannah Arendt's communal table: it "gathers us together and yet prevents our falling over each other, so to speak." (1998[1958], p. 52).

The present chapter embraces all these previously mentioned facets of public space, and in addition, it aims to discuss their specific role in urban adaptation processes. More specifically, it will be argued that through the application of effective adaptation measures in public spaces, communities are facilitated to comprehend, learn, engage and mobilize for climate action.

As argued in this chapter, the distinctiveness of urban territories as major centers of communication, commerce, culture and innovation is what may empower successful processes and outcomes of the climate change adaptation agenda. In addition, recalling Jordi Borja, the cities' interchange processes of products, services and ideas "need, are processed and expressed in their public spaces" (2003, p. 120, author's translation).

As argued by Banerjee, there has been an increasing tendency for people to gather in order to improve a common livability, often ending up enhancing the overall quality of the urban environment (Banerjee, 2001). Banerjee additionally evidences that most of these shared public actions happen in existing public spaces, such as streets, squares or school amphitheaters, therefore "reasserting the role and sustenance of the public realm" (Banerjee, 2001, p. 15). Not only do people want to be the main actors in the urban space, but they also want to be at the center of space design concerns. As Jane Jacobs pointed out,

> Dull, inert cities, it is true, do contain the seeds of their own destruction and little else (...) lively, diverse, intense cities contain the seeds of their own regeneration, with energy enough to carry over for problems and needs outside themselves.
>
> (Jacobs, 1992[1961], p. 448)

Indeed, what is regularly overlooked in large-scale planning and policy – often guided by questionable interests – is a community's inherent resilience (Figure 1.2).

Figure 1.2 Left: Children of Restelo neighbourhood facing the warm summer of Lisbon, taking a bath in an inflatable pool placed near their house yet outside in the public space. Right: Group of young people having fun by interacting with water on a summer day at Cais das Colunas, Lisbon, Portugal.

Image credits: Soraia Noorali, Agosto, 2016 (left); Author's personal archive, 2011 (right).

Climatic hazards such as flooding, potentially aggravated by climate change, are an increasing threat that affects all the people in a community, particularly the most vulnerable (elderly, children and the poor, among others). Considering public space as a communal space, a collective entity of shared concerns, a new claim for climate change adaptation is presented: the claim that public space may therefore additionally serve as a social beacon for change.

In light with Pelling's findings, people and communities are not only targets but also active agents in the management of vulnerability (Pelling, 1997). Ulrich Beck also highlighted that "what was made by people can also be changed by people" (Beck, 1992, p. 157). Correspondingly, not only is it in the public space where hazards become tangible to a community, but it may also be where adaptation initiatives may strive. It therefore comes as no surprise that a new variety of insurgent citizenship is arising within public spaces as the urgent matter of climate change adaptation is recognized among our societies.

Recalling just two among other insurgent examples, recall the provocative phrase Banksy wrote in the front of a house that flanks the Regent Canal, north of London: "I DO NOT BELIEVE IN" and the bottom line "GLOBAL WARMING" cut horizontally in half as if suggesting that the water level had already risen; or the phrase "WE ALL HATE YOU SANDY," probably impulsively written on the wall of one of the many houses devastated by this hurricane in New York, 2012. Regardless, bearing in mind the potential severity of the projected impacts that are expected to become increasingly more unavoidable, many authors agree that our society is still not responding accordingly (IPCC, 2014a).

Some societies have shown to be reluctant of the need to face impending threats of climate change. It seems as if there is no common understanding of what is the "common good." Hesitant communities may be driven by the fact that climate change is still a much-politicized issue or by the fact that adaptation is still a fairly recent strategy of response. Regardless of the causes in climate change suspicions, some cases evidence an inclination to prioritize other values.

Even within developed countries that have already suffered direct consequences of severe climate impacts, some communities have rejected initiatives toward a more adapted urban environment. That is namely the case of the New Orleans local society, which reacted against the construction of an adaptation plan proposed by a group of experts after the Katrina incident of 2005:

> On a Friday morning in ravaged New Orleans, Louisiana, Joe Brown learned just how fiercely people value their homes. Along with several dozen other disaster experts, the veteran urban planner had been recruited by Urban Land Institute in Washington D.C., to develop a rebuilding plan for the city, which had been devastated by Hurricane Katrina in August 2005 (…) about a quarter of the city lay in utter ruin and remained at high risk of flooding. Brown displayed diagrams that suggested turning some blocks, for the time being, into open space.
>
> Reaction was swift and harsh. A council member accused Brown of aiming to "replace these fine neighborhoods with fishes and animals" he recalls. A couple of audience members rose up and declared, "All we want to do is get back our homes." The planners were startled. "We got shock and amazement to what, to us, were fairly obvious truths," Brown says.
>
> (Couzin, 2008, p. 748)

Regardless that the practice of several construction techniques, such as flood resilient buildings, was met by a distinguishable positive impulse within local society, through the experience of Joe Brown, Couzin alerts the scientific and professional community about the various possible disparities between science and social understanding.

Other communities, in other situations, also did not initially welcome adaptation actions. That is namely the case of the first attempt to implement the currently internationally recognized concept of the water plaza, which can be briefly characterized by being a low-lying square that is submerged only during storm events. Despite a promising start – with an idea that, within the report "Rotterdam Water City 2005," had won the first prize of the 2005 Rotterdam Biennale competition – the first pilot project failed. Conflicts emerged from several sources, from prior conflicts with the municipality, which decided to conduct the pilot project, to the uncertainties associated to an experimental project of this nature. Risks, such as of children drowning, triggered strong emotional reactions from local citizens, who started naming the idea as the "drowning square" (Biesbroek, 2014, pp. 121, 122).

Having wisely learned from the encountered barriers that prevented the implementation of the first project, the second pilot project had not only a new location but, more importantly, a new approach toward technical criteria and social participation. This second attempt was successfully built and is currently considered an exemplary case of concrete climate change adaptation in a highly urbanized area.

Both mentioned cases evidence that social, cultural and emotional factors can be more valued and respected than the need of physical safety or ecological services of public spaces, a fact that strengthens the importance of continual community involvement alongside additional and distinct methods for the dissemination of scientific knowledge within the agenda of climatic adaptation.

According to Van Der Linden, persuasive communication about climate change is only successful when based on an integrated acknowledgment of the psychological processes that control pro-environmental behavior (2014, p. 274). In order to reach this goal, the author specifically claims three criteria need to be met: (1) the need to combine and integrate cognitive-analytical reasoning (knowledge/information), experiential (affective processing) and social-normative aspects of human behavior in the design of a message; (2) the need to make the climate change context explicit and (3) the need to target specific behaviors and their psychological determinants that need to be changed (Van Der Linden, 2014).

The role of public space as a mediator for social proactivity toward adaptation seems to have been insufficiently addressed in overall climate change discourse. In the cases where the local scale is considered, approaches often disregard context-specific features of communities and neighborhoods, to which people more likely connect, and are rather more oriented toward the policy role of municipalities. Yet public spaces seem to offer what Van Der Linden considered as fundamental for climate change adaptation engagement. Through public spaces and public space design, local aspects of climate change can be made visible and thus meaningful for citizens and their livelihoods. In his keynote speech at "2010 Deltas in Times of Climate Change Conference," the design director for Arup Urban Design, Malcolm Smith, expressed similar concerns when stating the need to "make visible the invisible" in climate change adaptation designs (Smith, 2010).

Furthermore, public spaces provide a differentiated source of knowledge and information (besides the mainstreamed sources of science and media) that may be apprehended as an autonomous and independent process – a process that may count with direct learning experiences, based on deep-rooted traditional experience and know-how, in a public domain that is naturally subject to social control. In other words, public spaces may provide extended opportunities for experiential learning that are influenced by specific contexts and social pressures. Through a medium that is closer to people, "climate change literacy" may more likely endorse a common need for action and search for solutions. According to CABE Space (Commission for Architecture and the Built Environment), a leading advisor of the UK Design Council, the adaptation of cities to climate-driven threats is strongly dependent on "well-designed, flexible public spaces"(CABE, 2008, p. 2). Others believe that "the best way to predict the future is to design it" (Buckminster Fuller in Brinke *et al.*, 2010, p. 2).

When integrating local expertise as well as scientific and technical knowledge in a flexible and clear-cut design, public spaces are not only able to promote adaptation action and reduce risk of disaster but also improve awareness on climate change.

The physical and the social components combined make public spaces favored interfaces for adaptation action. In public spaces people may "be" as well as "become" both producers and managers of adaptation action. People may "be" producers and managers of adaptation through autonomous, individual or collective involvements – from personal or art manifestations (Figure 1.3) to community-based projects. And people may also "become" both producers and managers of adaptation when awareness is raised through the direct consequence of the formerly mentioned processes or through institutional

Figure 1.3 After Hurricane Sandy. Graffiti "tag" denouncing the cause of destruction: "Global Warming." Queens, NY, November 2012.

Source: (Shutterstock 2012)

endeavors such as the design of a particular public space, by the message of a public artwork or by informative signage. In this line of reasoning, the design of public space sees itself enhanced in the face of impending weather events, being here considered as determinant for the adaptation of urban territories when facing climate change.

Regarding the particular matter of flood adaptation, public spaces entail further specific connotations besides those previously explored. Their specific features will next be identified through an empirical analysis based on a gathered range of examples. Underlying this assessment lies the argument that public spaces support the new emerging tendency on urban flood management, where the precedent goal to effectively and rapidly avoid or convey stormflows is being gradually replaced by the goal to incorporate stormwater within the city and through the enhancement of the whole natural water cycle. In other words, that public space, comprised with flood adaptation measures, helps promote a change of paradigm for more flood-adapted cities that aim to reduce vulnerabilities while integrating environmental, social and economic concerns.

3   Potential advantages of applying flood adaptation measures in the design of public spaces

In light of the range of studied examples presented in the doctoral research that supports this book (Matos Silva, 2016), it was possible to identify particular benefits offered by pubic space itself on the way toward more adapted cities. Among the distinguishable characteristics present within the range of analyzed examples, potential benefits may specifically arise by (1) the favoring of interdisciplinary design; (2) the possibility to embrace multiple purposes, (3) the promotion of community awareness and engagement and interaction, (4) the comprehension within an extensive physical structure, (5) the possibility to expose and share value and (6) the opportunity to diversify and monitor flood risk.

Each of these features will be scrutinized in the following pages and further reinforced by the association with additional identified cases from the "Portfolio Screening" presented in Chapter 3. Although this interpretation has been disseminated (Matos Silva and Costa, 2018), additional considerations are here included in order to reinforce the key role of public space design in flood adaptation endeavors.

3.1   *Interdisciplinary design of public spaces*

Acknowledging that public space ethics concept may be interpreted as "it is of everyone," its design is therefore "not a matter of one sole profession, entity or interest group" (Brandão *et al.*, 2002, p. 19, author's translation). Likewise, Madanipour argues that public spaces should be created by different professionals from different disciplines of the built, natural and social environments or by any professional with multidisciplinary concerns and awareness (Madanipour, 1997[1972]). As Lefébvre acutely states, it is the "ultimate illusion: to consider the architects, urbanists or urban planners as experts in space, the greatest judges of spatiality"[2] (in Brandão, 2013, p. 30, author's translation).

Recalling Horacio Capel's argument, since the 19th century, the subject of urbanism has been excessively controlled by a fierce competition between engineers on the one side and architects on the other. While the first would define and design major infrastructures, the latter would define and design interventions in streets, buildings or green areas.

However, as Capel highlights, "all this should be at the service of social needs" (Capel, 2005, p. 92, author's translation). In other words, all urban endeavors should firstly account for people and communities, and if necessary, should be able to surpass professional interests and the limitations of specialized expertise.

Back in the 1870s, Frederick Law Olmsted designed Boston's Emerald Necklace with the goal of resolving engineering problems of drainage and flood control together with the fulfillment of the increasing social needs for leisure and recreation opportunities in a growing population. Simply put, Olmsted demonstrated that it was possible to integrate complex connections between natural and technical processes together with an improvement of the quality of life of the surrounding populations. For Cynthia Zaitzevsky, "Olmsted foresaw that such a comprehensive approach embraced planning, engineering and architecture and that, to bring the disciplines together to create the best solution, needed the unifying instincts of the new profession of landscape architecture" (2001, p. 43). Today, Emerald Necklace parks include land and water features, engineering structures, public buildings and ecological designs that are merged together in a rational and balanced design (see Figure 1.4 – left).

Further examples of interdisciplinary public space designs can be namely seen in the city of Barcelona. As previously mentioned, it was likely due to Barcelona's urban regeneration grounding ideals from the 1980s that, in the beginning of the 21st century, the city decided to integrate the infrastructural construction of underground reservoirs underneath different types of public spaces, an interdisciplinary approach that required multiple professional areas to share their expertise throughout all procedural planning stages. By contrast, other municipalities have chosen to solely focus on one technical discipline. As a result, similar infrastructures were designed as isolated monofunctional facilities fenced off from their surroundings. Through Barcelona's integrated approach, it was further possible to enclose parallel advantages from a grand urban intervention, namely the creation of more public spaces. Putting it simply, Barcelona turned the constraint of a required great drainage improvement into the opportunity to build more public spaces for its citizens, with all its potential subsequent side benefits.

Other successful public space, its citizens, with all its potential particularly known for its interdisciplinary design that further entailed multiple purposes, is Potsdamer Platz in Berlin. Situated in an important area of the city, near the Berliner Philharmonie and the Berlin State Library, Potsdamer Platz has an approximate area of 1.2 hectares. Its design, composed of a series of urban pools, reveals an integrated approach between ecological, aesthetical and civil-engineering functions. The large water features are supplied only by pluvial waters. In summer, water surfaces lower the ambient temperature and improve microclimates. Roofs from the surrounding buildings capture rainwater and store it in underground cisterns. The collected water is then used for topping up the pools, flushing toilets and for irrigating green areas (Pötz and Bleuzé, 2012, p. 181).

One can further evidence a growing tendency for interdisciplinary design, specifically when interventions consider the need for climate change adaptation. One of the most recent examples of an urban realm that was created in light of the disseminated projections is the Olympic Park, more precisely Queen Elizabeth Olympic Park in London (briefly described in Chapter 3 – "Portfolio Screening"). In accordance, the park's landscape design priorities included "Great amenity; Improved micro-climate; Biodiversity; Integrated water management; Energy generation; Resource management; Waste management and minimization; Local food production" (LLDC, 2016, p. 1), among others – priorities that involved the inclusion of additional and uncommon disciplines actively involved in the design process.

Solutions arising from interdisciplinary designs are very diverse and combine the use of a wide range of approaches, such as technical, social, economic and ecological, among others. With regard to adaptation there is much we can learn from our civilizational past, which has overcome other great turbulences. We must also humbly accept that the impending future will require new outsets and new paradigms, new ideas that will most likely arise from a common effort of multiple, shared, and applied expertise. Public spaces, as spaces that particularly favor interdisciplinary convergence, may serve to promote and explore technological reinventions or innovations – a continuing learning process that, in face of climate change, searches for new design solutions that increase adaptability and reduce vulnerability.

3.2 Public spaces of multiple purposes

The resulting combination of an interdisciplinary design that integrates flood adaptation functions with public space design generally offers many side purposes among other sectorial needs such as recreation, microclimatic melioration or energy use and efficiency. Likely, the more interdisciplinary the design is, the more adjacent functions the resulting public space will comprise.

Traditional drainage infrastructure, for instance, such as large-scale underground retention chambers disconnected from public space, is only useful in certain occasional times during the year, namely during heavy rainfall. In contrast, other source control measures, such as green walls, bioretention basins or rain gardens, when applied within public spaces, may not only serve their prime infrastructural function but may also serve to improve local environment and quality of life as well as vulnerability reduction and local awareness. The side benefits that result from reconfiguring drainage infrastructure within public space design thus generally gather recurring advantages all year long.

The "Uptown Circle," a roundabout in Normal, Illinois, is an example that may evidence this argument. It is a green water square in a roundabout that collects, stores and purifies stormwater runoff from the nearby streets. Besides the aesthetical and leisure characteristics, the water feature serves to mask surrounding traffic noise, while purified water is used to spray nearby streets and thus lessen heat stress (Pötz and Bleuzé, 2012). The square further serves as a meeting place situated near a multimodal transportation center and a children's museum.

The previously mentioned Queen Elizabeth Olympic Park in London is one other example of an interdisciplinary design that consequently embraced multiple purposes, namely, the included treatment process that turns Londoner's wastewater from an outfall sewer into water suitable for irrigation, flushing toilets, and as a coolant in the Park's energy center (EEA, 2012b).

In addition, some breakwaters or wave-breakers, such as the one existing at the Zona de Banys del Fòrum in Barcelona, are here understood as multipurpose public spaces, as they not only entail the infrastructural function to ease the power of waves but also include the possibility to be used as a sightseeing route – two encompassing functions that are additionally combined in a sculptural design that is aesthetically appealing (Figure 1.4 – right).

Figure 1.4 Left: Emerald Necklace (detail). Right: Wave-breakers designed by Beth Gali, architect, at the Zona de Banys del Fòrum in Barcelona (detail).

Source: Author's personal archive, 2011 (left); Sara Garcia, 26 April 2014 (right).

3.3   *Public space for public awareness and engagement*

Social and political engagements are a particularly important factor in the success of adaptation endeavors, especially when acknowledging that adaptation is a learning process of continued assessment. Moreover, community engagement may be reached through community involvement, emotional connection and a design that makes visible the invisible. When adaptation actions are applied within a public space, where design can make visible certain intangibles, endeavors are no longer an abstract phenomenon for people and communities. Community engagement practices in public space design and management gain, therefore, a new dimension.

Considering flood events, which are expected to increase considering future climatic extremes, the roughly intangible water cycle can be made visible through design, particularly through the design of public spaces that, considering their inherent values, provide the opportunity to draw people in and connect them with water and thus potentially raise awareness and overall engagement. Indeed, as it is for Ashley *et al.*, the challenge of appropriate drainage systems for a changing climate is as much sociological as it is technological (Ashley *et al.*, 2010).

The Benthemplein square (Figure 1.5) and Tanner Springs Park may serve as examples that corroborate the argument that public spaces are rarely "mute" and may serve to connect people with water. Both cases encompass the concept of a "water plaza" that intentionally unveils part of the urban water cycle dynamics for the citizens that use that public space. As mentioned in the Rotterdam Climate Proof report, "Water disarms and binds people. In adaptation projects in the city, citizens and different cultures come together. This can reinforce social ties and the sense of safety" (2009, p. 7).

Regarding the case of Tanner Springs Park in Portland, its design comprised the restoration of a wetland into the setting of an urban context. Inspired by the area's original natural state, the park is composed of a pond at its lowest point, to which rainwater from the surroundings is conveyed. The design therefore combined several objectives

Figure 1.5 Benthemplein square, Rotterdam, the Netherlands (details). On the left, alongside benches and flowerbeds, are channels that, through gravity, convey stormwater into the square-reservoirs of lower elevations. On the right is the biggest square-reservoir, which may also serve as a sports field and a stage, among other things.

Source: Author's personal archive, 14 June 2014.

among the fields of ecology, water management, art and participation. Some of its main characteristics include reintroduced groundwater, water features, appropriate vegetation and site-specific artwork that evidences the native flora and fauna from the former wetland.

In order to promote community engagement, it is further important to highlight the need to create places that people can value and connect emotionally to. Likewise, the success of community engagement processes is strongly related with the development and value of local identity. In this sense, the presence of water in urban design, and more specifically in the design of public spaces, has symbolic dimensions (emotional, aesthetic and cultural) that should not be overlooked.

One of the oldest representations of water is Genesis's description of the Garden of Eden and its four structuring rivers that give life to this mythical space (Bíblia Sagrada, 1991, Genesis 2:10–14). Yet, as we are all aware of, water is not only the source of life, but it is also a permanent threat. And so the fear of water is also tattooed in our civilizations worldwide. Genesis's flood narrative in the Bible is one of many flood myths found in our cultures.

Water's symbolic dimensions should therefore be enhanced in a public space design that aims to connect people with water. This exercise is particularly evident in the works of Atelier Dreiseitl, which is an office that refuses to use water for pure decoration. Through its designs, it rather advocates for water to be integrated with other systems and other functions, always bearing in mind the final purpose of aesthetic appreciation and public perception of the value of water as a resource.

Another way to promote community engagement on the urgent need to adapt our urban spaces in the face of climate change is through direct community involvement and interaction – specifically because adaptation is not a once in a lifetime project. On the contrary, adaptation processes and projects require ongoing collaborations and organization between and among government, institutions and its citizens.

Greenfield Elementary in Philadelphia is a good example of the fruitful results that may arise from collaborative design among stakeholders – more specifically, parents, teachers, students, school administrators, designers from Community Design Collaborative and the Philadelphia School District. More importantly, all planning stages of the project until its end result worked as a living laboratory that taught anyone who passed by about the overall features of environmental processes. The plan aimed to convert the school yard, used previously as a parking lot, into a green space with sustainable concerns. The improvements included the installation of a flood management system with indigenous vegetation, the removal of impervious pavement, a permeable recycled play surface and an agriculture zone, as well as solar shading. A stormwater bioretention area with a rain garden was also installed.

Communal management also occurred in New Orleans after the destruction and desolation of Hurricane Katrina. More precisely, an "extraordinary new level of civic and community engagement" (ISC, 2010, p. 61) helped the city toward recovery and rebuilding, through a process that retained a strong connection to the city's history while also looking forward in addressing future challenges such as climate change. In accordance with the report developed by the Institute for Sustainable Communities (ISC) in partnership with the Centre for Clean Air Policy, the community embraced the idea that the best approach to endure future climatic extremes is to become a greener city that, consequently, promotes safety and enriches attractiveness for business and residents (ISC, 2010, p. 61). One of the implemented projects aimed to transform the constraint of having more than 60,000 vacant lost lots in the city, transforming some of them into a network of urban farms and public gardens. It is further important to evidence that in New Orleans' recovery, governments' investments alone would have had a reduced impact. The city was able to recover, and it is able to take forward its strategic plans, because of a creative and energized community, because of public and private partnerships and because of a comprehensive cooperation among national and international experts (Dutch Dialogues, 2011).

3.4 Public space as an extensive physical structure and system

Reflecting upon the perception of public space as a structuring element of the urban form (Martin, 2007; Portas, 2003), additional reasons promptly lead to the further conclusion that these are particularly favorable places for the implementation of flood adaptation measures.

Public spaces have a fundamental role in city life, as they enable formal and environmental continuity, accessibility and legibility, contributing to the reinforcement of social and economic centralities (Pinto, 2015). In the series of lectures "O Urbano e a Urbanística ou os tempos das formas" (2012), Nuno Portas highlighted that, in the history of cities, public spaces are more durable than buildings; that buildings are stable elements but not durable elements; and that after public spaces, the most durable elements are the buildings that are transformed into monuments, i.e., transformed into public spaces. Indeed, public spaces are determinant elements in the form and identity of a city.

According to Borja, "The fact that public space is the determining element of the urban form is enough to attribute it the role of structurer of urbanism and, firstly, its urban fabric" (Borja, 2003, p. 137, author's translation). One can thus claim that public spaces are not only the means of social, economic and cultural dynamics but are also a physical structuring element of the urban fabric. A structuring network that is able to construct a "recognizable and lasting image of an individual unity, which arises from a system of complementary parts, as various and as unorganized as they may be" (Portas, 2003, p. 17, author's translation).

By conforming a structural network based on the local scale, public space offers a decentralized and expansive means to tackle flood management, an approach that strongly contrasts with the traditional method, which tends to be linear and centralized. This distinction, together with an assessment of exploratory nature regarding the adaptation measure's infrastructural efficiency, was particularly emphasized in the article "Urban Flood Adaptation through Public Space Retrofits: The Case of Lisbon (Portugal)" (Matos Silva and Costa, 2017).

Moreover, public spaces offer a network that not only supports the urban fabric but also connects its different urban spaces, from buildings and infrastructures to natural structures such as the ecological network. For Portas, this communicating network of public spaces "cannot be reduced to a simplistic addition of segments, unconnected streets, detached from the territories they cross, more or less urbanized" (Portas, 2003, p. 17, author's translation). In other words, public space must not be understood by its individual elements but rather as a "coherent structure that encompasses different territorial scales (from the neighbourhood to the metropolitan city)" (Pinto et al., 2011, p. 1). The same can be said about hydrographic basins. Indeed, one of the main causes of urban floods is related to the manipulation of natural watersheds through forced interruptions or divisions into smaller parts. These approaches do not consider the fact that their effective functioning is highly dependable on a system that is comprehensive by nature.

It is therefore equitable to conceive water systems equally incorporated within the network of public spaces. One can easily identify episodes where water systems networks have met with public spaces. However, most of the time, it is an event that is neither planned nor wanted. Considering, for instance, drainage overflows resulting from heavy rainfall. In this situation, stormwater generally flows along the next available spaces, generally "open" spaces and mostly public spaces. If this "encounter" could be looked upon through a different perspective, one that would capitalize from the inherent values of public space, the excess of water could be integrated within designs as an opportunity to potentiate a comprehensive adaptation in an extensive and decentralized network. If, for instance, Lisbon's municipal undertakings, such as the Lisbon municipality's public space rehabilitation programs of "Pavimentar Lisboa 2015–2020" or "Uma Praça em cada Bairro," would have considered including flood adaptation measures in their design, significant benefits could have been gained toward a more resilient city (Figure 1.6).

A representative example that takes advantage of the benefits offered by the extensive physical structure of public spaces is probably New York's Green Infrastructure plan, launched in 2010. In brief, this plan aimed to offer a more sustainable alternative to the conventional "gray" infrastructure by proposing integrated structure that combined solutions such as rooftop detention, green roofs, subsurface detention and infiltration, swales; street trees, permeable pavement, rain gardens and engineered wetlands, among others.

While New York's program is illustrated by the example of Elmhurst parking lot, many other examples fit within its overall approach. More specific examples are the bioretention planters on Ribblesdale Road in Nottingham, United Kingdom, the open drainage system in the Trabrennbahn Farmsen residential area in Hamburg, Germany, or the drainage systems of the Ecocity Augustenborg in Malmö, Sweden.

The bioretention planters of Ribblesdale Road in Nottingham were a pilot retrofit project of sustainable urban drainage. They were therefore created for its design and construction to be documented and evaluated in order to assess its comprehensive application. A total of

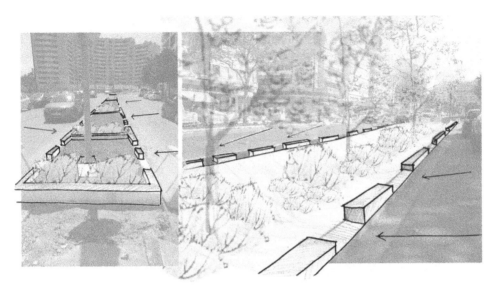

Figure 1.6 Rapid sketch of possible alternatives to the recent urban regeneration projects at Avenida Miguel Torga (left) and Avenida da Républica (right), at Lisbon (Matos Silva and Costa, 2018). The simple and inexpensive adaptation of the central road-separator's section would have improved the collection of rainwater and thus the attenuation of peak flows and amelioration of the urban water cycle.

Source: Author's sketch, 2017.

148 square meters of bioretention planters were implemented within an existing urban road setting. Among the main objectives of this intervention, the following are worth mentioning: (1) maximize surface water interception, attenuation and infiltration; (2) encourage participation from local residents in the design and future management of the rain gardens; and (3) evaluate the effectiveness of the scheme as an engagement tool around the sources of urban diffuse pollution and flood risk (Susdrain, 2019).

Trabrennbahn Farmsen is an example of a newly built residential area that comprised the application of a particularly interesting open drainage system. Because its implementation area has little infiltration capacity, designers chose to implement an open water system that would retain and convey rainwater. In accordance, stormwater is collected from surrounding streets as well as from the building's roofs. Overall the system is composed of grassed swales, stormwater channels and two retention ponds (Howe *et al.*, 2011, p. 66). The greatest highlight of this example is the autonomy of this natural system to manage all stormwater from the Trabrennbahn Farmsen residential area on-site, evidencing a reduced importance of underground sewers for rainwater.

3.5   *Expose and share value through public space*

By integrating infrastructure in the design of a public space instead of camouflaging it underground or in an isolated impenetrable area, a public investment is exposed and shared with a community, a shared value that may instigate further opportunities such as amenity or environmental quality. For example, while in the common mainstream urban drainage approach investments are camouflaged underground, frequently encompassing a sole function and used for stormwater alone, investments on urban drainage could be applied in infrastructure that is integrated within the public space itself. In the second option, value is not only exposed to all but also shared among everyone using that space. Sustainable urban drainage systems (SUDs) clearly illustrate and make the case.

Through dispersed yet extensive small-scale investments within public space design, urban amenities are further created while taking advantage of ecological and economic opportunities along the way. While buried culverts may be a missed opportunity for the enhancement of the quality of public space, obsolete and no longer necessary flood walls may likewise hide valuable water assets (Papacharalambous et al., 2013). Indeed, there are many opportunities for infrastructure renovation and necessary landscape improvements throughout urban areas: from the need to provide alternatives to reduce the load of obsolete drainage infrastructure to vacant lots that can be used to store water. Bearing in mind the ever burgeoning costs of traditional infrastructure repair and expected climate change extreme events, a wide range of literature argues that established methods are no longer affordable or sustainable (such as White and Howe, 2004; Hartmann and Driessen, 2013; Lennon et al., 2014). As such, new alternatives, which instead support an integrated water management, should be considered not only as a necessary immediate investment but, more importantly, as an investment in our future.

Uncovering small scaled stormwater drainage systems, such as in Banyoles, Girona (briefly described in Chapter 3 – "Portfolio Screening") or maintaining traditional infrastructure, such as in the 13th century Freiburg Bächle, is one way of exposing and sharing the expressed value of water in an urban environment.

Projects that "bring into light" buried preexisting water lines are another example of an adaptation measure that aims to expose and share value. That is the case of Westersingel channel and the Soestbach river at Soest, besides the representative example of the Cheonggyecheon river (briefly described in Chapter 3 – "Portfolio Screening").

Rotterdam's Westersingel channel, which had been previously sunken, was redesigned by Dirk van Peijpe from the De Urbanisten office. The resulting promenade almost disguises its capacity to sustain and retain excesses of water when needed (Figure 1.7 – left). Its banks are mostly made of brick as well as grass and trees. All materials, including urban equipment such as benches and lamps, are designed to endure occasional overloads of the canal. Currently this public space is enriched with sculptures by well-known artists such as Rodin, Carel Visser, Joel Shapiro and Umberto Mastroianni.

3.6   *Public spaces as a means to diversify and monitor flood risk*

According to the IPCC, "The main challenge for local adaptation to climate extremes, is to apply a balanced portfolio of approaches as a one-size-fits-all strategy may prove limiting for some places and stakeholders" (Intergovernmental Panel on Climate Change (IPCC), 2012, p. 291). In other words, considering climate change, it is not recommended the sole investment in one isolated infrastructure, built to fulfill only its particular purpose. If plan A fails, risk will be great and generalized. But if investments are diversified, risk is dissipated through the reliance on parallel plans.

In addition, when massive infrastructures are kept out of sight, people do not remember their existence and thus will not expect their failure. This unpreparedness, led by a false sense of safety that is usually termed "levee effect," may further exacerbate vulnerability and amplify potential impacts. Contrastingly, if approaches are implemented within public spaces, some risks are more closely acknowledged and thus less unexpected.

Regardless, research aimed at analyzing the social construction of risk or social risk perception is rather complex. As evidenced by Sergi Valera (2001), social theories of risk suggest that the causes and consequences of risk are mediated by the subjective criteria of individual (or psychological) processes, social (psychosocial) behaviors and culture. The same way the design of a public space may reduce or exacerbate a risk through "rational and scientific" processes, it may also reduce or exacerbate the perception of that risk through "subjective-social" processes. It is further important to note that the social construction of a risk may influence the degree of the risk itself, minimizing or maximizing it. Risk perception is thus a very important factor that must be taken into consideration, particularly in the design process of a public space with flood adaptation purposes, so that the resulting outcome does not contradict the initial purpose.

Through the diversification of risk, by investing in more than one great monofunctional strategy, the communal management among government, institutions, communities and private companies is also promoted, unlike traditional management, which is essentially based on government's actions on behalf of communities. As a result of communal management, risk is further shared, and communities are more likely involved in the management and monitoring of implemented infrastructures. In addition, local know-how is explored, and citizens are empowered to act before the need for safety and the identity of vulnerable places is reinforced.

In a way similar to the representative example of Blackpool (briefly described in Chapter 3 – "Portfolio Screening"), Scheveningen's waterfront boulevard also illustrates the difference between its previous coastal defensive approach and the outcomes of the new implemented solutions. Designed by Manuel de Solà-Morales, this former project demonstrates how a large-scale storm surge defense infrastructure, such as a "super levee," may be integrated within the design of a multifunctional public space. Comprised by three levels of parallel and undulating waterfront boulevard, this space offers more than its functional requirement to protect The Hague from coastal floods. More specifically, it articulates other programs such as coastal life and recreation (bars and restaurants), public and private circulation (bicycles and cars) and connections with the urban fabric (Solà-Morales, 2012).

One may further consider the example of Ribeira das Naus, also a waterfront urban regeneration project in Lisbon's historic center, designed in partnership between PROAP and Global landscape architecture offices (Figure 1.7 – right). In this design, the monotonous embankment that accompanies most of Lisbon's waterfront is transformed into a ramp with steps that calls passersby to approach the river, an apparently simple design decision that brings people to more closely acknowledge tide cycles alongside other water-related natural occurrences such as storm surges.

Indeed, through these projects, people are more connected to the sometimes intense coastal water dynamics and thus more aware of its nature. By sharing the value of the infrastructure through its common use as a public space, not only is the awareness of the power of nature promoted but also a certain sense of responsibility and appropriation is reinforced.

Figure 1.7 Left: Rotterdam's Westersingel channel, the Netherlands. Right: Ribeira das Naus, Lisbon, Portugal.

Source: Author's personal archive, 2011 (left) and 2014 (right).

While the first aspect may lead to the respect and willingness to adapt, the second aspect may lead to active management and monitoring of the infrastructure itself.

Discussion

Within the multi-scaled scope of adaptation action, local scale "from the bottom-up" adaptation is particularly relevant, not only because it very likely influences global climate but also because it entails immediate repercussions for the reduction of society's vulnerability. Not only are hazards more acutely felt at the local level, but it is also within local communities that most know-how and experience are found to promptly deal with existing vulnerabilities. As highlighted in this chapter, competent and politically autonomous municipalities that are close with their citizens are therefore more likely to conduct effective adaptation action with wide-ranging positive repercussions.

Moreover, as evidenced particularly by the Barcelona model, local initiatives, such as specific public space interventions, raise the standard for good quality cities. Likewise, and as argued by several authors, the quality of cities can also be measured by the quality of its public spaces (Borja, 2003; Brandão, 2011a, among others). In this line of reasoning and bearing in mind the specific advantages of local-scale adaptation, it is argued that the quality of our future cities will be influenced by the quality of future adaptation measures applied in public spaces. Indeed, public space enables and promotes community life. Public spaces further potentially offer wide-ranging benefits such as placemaking, sense of place or local identity. As a civic common space, a collective entity of shared concerns, a new variety of insurgent citizenship is arising within public spaces as the urgent matter of climate change adaptation is recognized among our societies. Furthermore, as previously emphasized, social, cultural, and emotional factors can be more valued and respected within a community than the need of physical safety or ecological services. As such, through public spaces and public space design, local climate change can be made visible and consequently meaningful for citizens and their livelihoods. Public spaces additionally provide a different source of knowledge and information, besides the main-streamed sources of science and media, which may be apprehended as an autonomous and independent process. Accordingly, public spaces provide extended opportunities for experiential learning inherent to adaptation processes.

In this line of reasoning, the design of public space sees itself enhanced in the face of impending weather events, being considered as a key factor in the adaptation of urban ter-ritories when facing climate change and flood events in particular. Indeed, besides provid-ing the means to include flood adaptation features, public spaces per se entail further specific connotations that are advantageous in adaptation endeavors; potential benefits may specifically arise from the characteristics of public space to:

- Favor interdisciplinary design – in places founded through interdisciplinary means, innovative thinking more easily emerges;
- Embrace multiple purposes – by combining flood adaptation measures with public space design, adjacent purposes arise among other sectorial needs such as water depuration, recreation, or microclimatic melioration;
- Promote community awareness and engagement – by engaging the community in the design and use of a public space, not only awareness about climate change may be promoted but also the self-determining willingness for adaptation action is enhanced;
- Be supported by an extensive physical structure and system – by conforming a com-municating structural network, public space offers the advantage of a decentralized and expansive means to tackle flood management;
- Expose and share value – by integrating flood management infrastructure in a public space design, instead of camouflaging it underground or in an isolated impenetrable area, a public investment is exposed and shared, a shared value that may instigate further opportunities such as amenity or environmental improvements; and
- Promote risk diversification and communal monitoring – by investing in flood adaption measures applied in public space in addition to the conventional approaches, risk is dissipated and diversified and thus reduced. Moreover, through the diversification of risk, communal management among varied stakeholders is promoted. This way, communities are more involved, and the sense of responsibility and appropriation is stimulated, thus potentially leading to autonomous management and monitoring of shared implemented infrastructures.

People and communities can thus be perceived as more than susceptible targets and rather be professed as active agents in the process of managing urban vulnerability; climate change literacy, through the design of a public space, may endorse an increased common need for action and the pursuit of suitable solutions; and local know-how and locally driven design can be considered as an added value for adaptation endeavors.

Yet it is important to bear in mind that, in light of an unprecedented area of action, con-cepts, paradigms or structures are expected to change over time, as are the functions, appearance and complexities of public spaces with flood adaptation measures. The design of these spaces must therefore encompass an ongoing process that is fundamentally grounded on the need to learn, reflecting upon mistakes and generating experience while dealing with change (Berkes *et al.*, 2003). In the words of Jordi Borja, today we must "Accept the challenges with the intent to provide answers and with the modesty of providing them with uncertainty, with the audacity to experiment and with the humility to admit mistakes" (2003, p. 140).

It is furthermore essential to highlight that while the analyzed initiatives have counterbalanced the inevitable uncertainties of global models and the generalized "top-down" policies, local action must be connected to the global scientific findings and its encompassing strategies. Otherwise, applied adaptation measures may get lost in scale and lose their value and thus fail their purpose. While local-scale action is presently acknowledged as a fundamental element for effective urban climate change adaptation, its greater challenge therefore relies on finding the balance and explore the benefits from the arising synergies between local collective actions and national and international strategies. The same way local adaptation strategies must not be dissociated from global adaptation strategies, so too do the processes of public space design, which must follow objectives and strategies of regional and national levels, otherwise "actions will not contribute, in an effective way, to the achievement of community expectations in the safeguarding public interests and collective resources" (Brandão *et al.*, 2002, p. 19, author's translation).

Through the inclusion of flood adaptation measures within public space design, new challenges arise before contemporary urbanism and urban design practices. Likewise, although this research is specifically focused upon adaptation measures applied in public spaces, there are several other areas of opportunity that may additionally provide significant contributions in the development of flood-adapted cities. More specifically, disciplines such as building design, governance or landscape architecture have been extending their literature regarding this specific subject matter, suggesting further developments, especially of floating buildings, transdisciplinary and transboundary consortiums or in blue and green corridors.

Overall, sustained by an empirical analysis based on specific examples (Matos Silva, 2016; Matos Silva and Costa, 2018), this chapter aims to emphasize the specific advantages offered by public space as a means where flood adaptation measures can be implemented. However, it is important to note that the presented findings are not intended to serve as restrictive boundaries. It is not here advocated that flood adaptation endeavors can only be considered as such if comprising all the mentioned potential advantages offered by public space, nor that they are only successful if comprising all these mentioned advantages. What is argued is that the aforementioned characteristics are only potential and are considered as an additional asset either alone or combined. Ultimately, it is reasoned that public space is an ideal interface for adaptation action. Consequently, it is further questioned whether the evaluation of adaptation initiatives should consider: (1) if the design of a public space comprises adaptation measures and, on a reverse perspective, (2) if the application of adaptation measures comprises the design of a public space.

Nuno Portas (in Brandão and Remesar, 2003) reflected about different phases of urban projects that have led to different ways how public spaces have been produced. In the described first phase, most interventions were held in heritage areas, entailing projects such as the pedestrianization of historic centers or creation of public spaces as a replacement of old industrial uses. The second phase entailed urban projects that were induced by events such as the Olympic Games, Capital of Culture or international exhibitions. These projects had in common the aim to generate new facilities suited for leisure, culture or sports as flag/brand attracting projects. Proposing a further prospective discussion, a final question arises: are we at the fringe of a third phase, in which, in a changing climate, urban projects of diverse territorial nature will also aim to produce public spaces that are prepared to adapt to future impending weather events?

Notes

1   Among the most recent public demonstrations supporting the imperative need for climate change response is the papal encyclical letter "Laudato Si.'" Throughout his letter, Pope Francis appeals for a renewed discussion about the future of our planet, particularly considering that "Never have we so hurt and mistreated our common home as we have in the last two hundred years" (2015, p. 39). Although recognizing the lack of culture needed to confront this crisis, Pope Francis does not give up hope that "humanity at the dawn of the twenty-first century will be remembered for having generously shouldered its grave responsibilities" (2015, p. 123). In order for this change to happen, social vices for comfort and superficiality, which tendentiously assume that nothing serious will happen, must be inverted.

Counterbalancing this tendency, the 2015 United Nations Climate Change Conference (COP 21 Paris) marked a new threshold for climate action. This was an important undertaking, particularly considering the number of parties involved. More specifically, 195 world nations have agreed to sign the so-called Paris Agreement, as opposed to the 114 parties of the previous Copenhagen accord. The only countries that did not accept the agreement are North Korea and Syria, for understandable reasons. It is therefore considered a legal universal agreement against climate change, with great prospects to effectively reduce or eliminate the consumption of coal, oil, and gas as energy sources that have fed human societies since the 18th century.

2   Original text: "*Suprême illusion: considerer les architectes, urbanistes ou planificateurs comme experts en espace, juges suprêmes de la spatialité.*"

Chapter 2

Categories and types of flood adaptation measures applicable in the design of public spaces

Introduction

A state-of-the-art review on previously developed frameworks on climate change adaptation measures supported the initial identification of the existing categories and types of flood adaptation measures applicable in the design of public spaces. This research was particularly targeted to external research findings that, directly or indirectly, studied the subject of flood adaptation measures. Details regarding the research methods, results and discussion have been previously published (in Matos Silva and Costa, 2016).

Previous research findings have identified, described and characterized different types of adaptation measures in accordance with the scope of their studies. In most cases, the exposed definitions entail specific purposes or approaches directly influenced by the measures' execution technics, materials, costs or the assessment of their infrastructural efficiency. In the same line of reasoning, yet in coherence with the thesis presented in the previous chapter, the following text will aim to identify and briefly characterize the different types of flood adaptation measures that can be specifically included in the design of public spaces.

A detailed technical characterization of each measure is unintended, as that work has been previously developed in numerous references (Birgelen *et al.*, 2011; Harris *et al.*, 1998). What is here intended is to evidence, through a straightforward vocabulary, the numerous design options and their basic attributes for the adaptation of public spaces to flood events. Ultimately, the goal consists of moving out from the intricacies of specific disciplinary areas into the promotion of these adaptation measures as a mainstreamed design practice. The following descriptions of each identified measure will therefore dwell upon each respective practical and operational considerations, bearing in mind the implications in the design of a public space. Whenever appropriate, specific attention is given to their primary and secondary infrastructural function,[1] the materials that can be used in their implementation, the type of flood to which they are more appropriate (Figure 2.1), the scale and extent of their benefits and their fundamental maintenance requirements, among others.

The presented descriptions are mainly supported by the literature from the previously highlighted state of the art overview (in Matos Silva and Costa, 2016), specifically Philip (2011) and Prominski *et al.* (2012), together with other consulted references such as Novotny *et al.* (2010) and LNEC (1983). As such, based on a scientific, gray and technical literature review, as well as on data collection, networking, site visits and empirical observations, a synthesized characterization of each category and type of measure is then proposed. Drawn schemes of each type of measure were further developed, not

Figure 2.1 Floods are generally typified according to their cause or source. They can be divided into (1) fluvial, (2) pluvial, (3) coastal, 4) groundwater or (5) artificial drainage floods. Other authors further evidence the type of floods associated to (6) glacial lake outbursts. In addition, severe fluvial floods and costal floods can push man-made flood defense infrastructure, where they exist, up to unsafe thresholds (1a, 3a). When costal defenses rupture, fluvial and costal floods are significantly exacerbated. For more information regarding each type of flood, please consult Matos Silva (2016).

Source: Author's design, 2015.

only as a tool that helped the reasoning process behind the proposed typification and conceptualization but also as facilitating graphic information. In the end of this characterization section, all the sketched typologies are presented in one single matrix.

1   Urban greenery

Every living plant can be considered as part of the urban greenery of a city. From the smallest bryophyte (moss) to the biggest tree, all living vegetation contributes to the "harvest" and "detention" of rainwater, thus contributing to the adaptation of floods specifically associated with rainstorms (pluvial floods). Measures within this category further include adjacent benefits such as carbon capture, microclimatic balance or biodiversity enrichment, both flora- and fauna-wise. Although trees and bushes (whether singularly or in groups of random or aligned forms) form part of the category of "urban greenery," "green walls" are the only highlighted measure, essentially because of its increasing use as an adaptation measure with significant aesthetical and infrastructural contributions (Kew *et al.*, 2014).

1.1   *Green walls*

Green walls, in the exterior, are widely known for their functional qualities of thermal regulation, biodiversity and rainwater harvesting. Green walls can also be known as vertical gardens, living walls or plant walls. They are seen in both public and private properties. Most examples are an integral part of or influence its encompassing public spaces.

There is no minimum space required to install a green wall. They are usually installed in blind façades, although they can be applied in almost any type of wall, particularly walls with different materials or with varied available space. In order to maintain the high performance of green walls, a high investment in maintenance is fundamental, particularly in irrigation and drainage systems, as well as for continued checkups for pruning, deadheading and weeding.

There are many ways to design green walls. One can opt to use climbing vegetation, knowing that different species reach different heights (*Wisteria sinensis* may reach 25 m height, *Parthenocissus tricuspidata* may reach 20 m height, *Vitis spec.* may reach 15 m

Figure 2.2 Green wall in Campolide, Lisbon, Portugal. From left to right: general view; detailed view of "soil-in-a-bag" technique; and detailed view of metal supporting vertical structure detached from the building wall.

Source: Author's personal archive, 8 February 2019.

height, *Clematis vitalba* may reach 10 m height, etc.). Another way of designing green walls has been patented by Patrick Blanc, who called it *mur vegetal*. Climbing vegetation is not usually used in this technique, but rather a varied number of small-sized species that can survive steep slopes. Indeed, this approach encompasses a vastly different technology, especially regarding the supporting medium, which substitutes the mechanical aspects of common soil with light plastic elements mounted in a vertical frame. Water and fertilization are made automatically in installed systems.

Most commonly, green walls are built over panels that sustain a growing medium. Among the different types of support media are loose media; mat media, which can be of either coir fiber or felt mats; and sheet media, usually of polyurethane and structural media that combines loose or mat media in a structural block that can be designed with various forms (Figure 2.2). In other situations, vegetation emerges from ground planters, especially when using climbing vegetation. In this situation, green walls can be built directly over a wall or in a supporting structure. When green walls are built directly over walls, special attention must be given to the construction details in order to avoid moisture problems by contact or condensation, either in the base of the building or in its supporting wall.

There are many iconic examples of green walls in most major worldwide capitals. The green wall of Westblaak car park silo in Rotterdam exemplifies the impact that this measure can have when extensively applied in an urban element, such as an urban block. Although it is still far from reaching this goal, it aims to fulfill the 5000 m² of façade surface, equaling 200 mature trees. Rainwater is collected by the green structure and is used to irrigate the structure's own plants, thus relieving overburdened sewer systems. It will further enrich the surrounding urban environment, both aesthetically and ecologically, through the enhancement of the city center's biodiversity.

2   Urban furniture

Urban furniture as a contributing element to flood adaptation started to encompass innovative ideas, particularly when facing the challenges presented by climate change. Among the materialized ideas are the "inverted umbrellas" and "art installations" for specific purposes.

2.1   Inverted umbrellas

As the name indicates, inverted umbrellas resemble upside-down traditional umbrellas and may have various shapes and sizes. With inverted umbrellas, incoming water flows inward into the central column, which may then lead it to a storage tank or into the drainage system for effective reuse or distribution of rainwater. Besides taking advantage of rainwater as a precious and finite resource through harvest and storing functions, this measure may also serve as a shadowing structure, thus further contributing to microclimatic melioration through albedo reduction. This measure is particularly suited to adaptation to pluvial floods. Examples of inverted umbrellas include the mega-water-collecting-structure at the boulevard of Expo Shanghai or the small inverted umbrella applied in a playground at the Woolworths store in Walkerville, Australia.

2.2   Art installations

Most art installations, particularly the ones associated to flood adaptation purposes, have as their primary motive the communication of information, namely of a problem or concern, and thus are more likely be related to strategic measures such as warning and escape information instead of operational design measures. Yet in some cases, art installations may also involve operational measures encompassing a specific function. That is the case of the art installation at Jawaharlal Planetarium Park in India, or the art installations of Buster Simpson such as the Water Table/Water Glass and the Whole Flow.

The Water Table/Water Glass intervention, located at Ellington Condominiums square in Seattle, Washington, represents a metaphor composed by two elements, "glass" and "table," which are both fountains. Both fountains are fed by the rainwater collected in the ten-story towers' roofs: the "glass" is filled by the south tower, while the fountain embedded within the "table" is fed by the north tower. The 2.4-meter-high tapered "vessel-glass" is also a container that serves as a detention tank. Collected rainwater is expected to be of use for irrigation purposes (Buster Simpson, 2015).

The Whole Flow intervention, another example briefly explained, is an artistic solution for a downspout. The designed structure recovers rainwater from the roof, aerates this collected rainwater through a number of flowing plates, and then finally directs the (now recycled) water into a plant sand (Buster Simpson, 2015). The process of water descending through the bowls and cleansing its way onto being useful for irritation may be seen and heard by whoever crosses by the Whole Foods Store in Pasadena, California.

3   Rooftop detention

Roofs are privileged spaces among urban territories, as they encompass vast spaces from which water can be easily harvested and collected for later use. Rooftop detention systems mostly comprise a multi-layered structure that is designed in accordance to the function and size of the roof system. This measure can be applied in small- or large-scale spaces, from singular buildings to housing developments, from elevated obsolete train-lines to industrial estates. Public spaces may exist among any of these areas.

Besides the fact that capturing and reusing rainfall from roof surfaces can attenuate overall urban runoff, rooftop detention further contributes as a reliable source of water for irrigation and other non-potable uses on-site, such as toilet flushing, heating systems and direct groundwater recharge, among others. Commonly, the category of "rooftop detention" is associated or integrated to other identified categories and their respective types of measures, such as "reservoirs," "bioretention" or "permeable pavement."

The identified measures encompassed within the category of rooftop detention are "green roofs" and "blue roofs," both of which are particularly suited to adaptation to pluvial floods.

3.1   Green roofs

As the name suggests, green roofs are generally characterized by roofs covered with vegetation. Green roofs are usually divided by their range of vegetation "intensity." This characterizing range of intensity goes from green roofs of intensive use to green roofs of extensive use. Extensive green roofs have thin *substratum* and have few, usually succulent, plants. Intensive green roofs include a thick soil medium in order to sustain deep-rooted vegetation. Intensive or extensive green roofs can be applied in both plain and sloping surfaces, although special attention must be given to the details of its design so that the roof remains water-hermetic.

Rainwater that is harvested by green roofs is purified by natural processes of its composing vegetation and can later be let into other storage devices such as tanks or cisterns in order to be reused.

On a large scale, green roofs can help mitigate the heat island effect and contribute for the reduction of overall urban runoff. Green roofs may further improve thermal and acoustic insulation of buildings, promoting the reduction of energy consumption for heating and cooling. In the winter, green roofs contribute to stabilize the temperature of the living space, serving as an additional heat insulating layer. During the summer, evaporated soil moisture resulting from the incidence of solar radiation provides a cooling effect that benefits all adjoining areas.

There are many examples of public space green roofs. Among the analyzed cases in the scope of this research are the Museum of Art, Architecture and Technology (MAAT) and Alcântara Wastewater Treatment Plant, both in Lisbon, Portugal (Figure 2.3), or the

Figure 2.3 Left: Green roof public space at the Museum of Art, Architecture and Technology (MAAT), Lisbon, Portugal. Right: Green roof at Alcântara Wastewater Treatment Plant, Lisbon, Portugal.

Source: Author's personal archive, 24 June 2019 (left) and 22 May 2015 (right).

Figure 2.4 Left: Detail of the "Promenade Plantée" in Paris, France. Right: Detail of the "High Line Park."

Source: (ChristopherGeorge 2018) (left); Source: Author's personal archive, 1 April 2019 (right).

Ewha Woman's University in Seoul, South Korea, both of which emphasize how a green roof can be specifically designed to sustain a public space. Other examples also use this technique in order to retrofit preexisting infrastructural transit viaducts into public walkways, such as the "Promenade Plantée" in Paris, France, or the "High Line Park" in New York, US (Figure 2.4).

3.2 *Blue roofs*

Blue roofs, also known as water roofs, are similar to green roofs except that they use various types of flow controls to regulate, block, or store water instead of vegetation. More specifically, blue roofs commonly use downspout valves, gutter storage systems, and cisterns (CCAP, 2011, p. 11).

Esplanades, for example, can be built over blue roofs, through a floating or elevated structure, or even placed over a permeable pavement connected to a cistern. Blue roofs may further serve as a water feature to be appreciated by the surrounding public spaces, as may be evidenced by the case of the Walterbos Complex at Apeldoorn, the Netherlands.

Once rainwater is harvested by blue roofs, it may be diverted onto adjacent public spaces and different types of adaptation measures. This complementary situation between blue roofs and other measures can be illustrated by the Stephen Epler Hall case at Portland State University, where rainwater is collected in blue roofs in order to be conveyed into rain garden planters and, subsequently, a cistern at the ground floor.

4 Reservoirs

In the scope of this analysis, the category of "reservoirs" groups measures that store water either from traditional "gray" sewage drainage systems or from other systems using measures such as bioswales, blue roofs, or green walls. Their main common purpose is not to lessen the volume of urban stormwater through natural infiltration or evapotranspiration but to provide temporary storage during heavy rainfall and thus attenuate intensity peak flows through the posterior gradual release of runoff. Accordingly, they are rarely stand-alone solutions, requiring supporting infrastructure both up- and downstream of their implementation site. Among the measures related to this category are "artificial detention basins," "water plazas," "underground reservoirs," and "cisterns."

4.1 Artificial detention basins

There is semantic difference between "retention" and "detention" that must be highlighted here. In brief, in retention systems water is generally harvested and reused; it stays on site and infiltrates into the soil. In detention systems, water is intended to slowly drain off-site through streams or drain pipes. Detention basins may remain ponds in places with a high water table. Unlike in bioretention, almost all the water from detention basins keeps flowing onward (Forman, 2014), regardless of the evaporation, infiltration and sedimentation of solids that may insignificantly occur. Retention systems are more suitable when subsoil percolation is reliable, whereas detention systems are more suitable when subsoil percolation is not reliable (Urban, 2008). In accordance, detention basins may be artificial or naturalized as long as they fulfill their primary purpose to solely detain and slow stormwater rather than also permit its infiltration.

Specifically, artificial detention basins are aboveground impermeable reservoirs, usually located in downstream areas of lower topographical elevation. These basins, constructed with building materials, receive rainwater from their surrounding area or from a distance through drainage systems that separate rainwater from gray-water. Artificial detention basins resemble sealed impermeable lakes, albeit usually with a variable water level. In some situations, these systems entail separate storage tanks in order to facilitate the control of water retention and subsequent continuing slow evacuation.

It is important to note that stagnant water can be very dangerous to public health; therefore, this measure should always consider the installation of water features such as fountains or cascades, promoting water dynamics and consequent water oxygenation.

This type of measure can be built in various shapes and sizes, from a simple rectangle, as illustrated by the example of Poblenou Park, to an undefined form of strong edges, as illustrated by Diagonal Mar Park (both cases within the city of Barcelona).

Recalling the example of the Baray reservoir in the Temple of Preah Vihear, this type of measure can be considered as an ancient technique of water harvesting and storage for the latter irrigation of farmland, further encompassing a careful and appealing composition. Also, the many examples of tanks existing within the Portuguese traditional farms with formal gardens (*Quintas de Recreio*) are said to serve similar aesthetic and production purposes.

4.2 Water plazas

The water plaza or water square can be generally characterized by being a low-lying urban area with a configuration similar to a square, which can be submerged during storm events. It is a measure that fits particularly well in densely built-up urban areas and not only in downstream areas or areas with low topographical elevation. It is designed to collect and retain rainwater from the nearby surrounding area, arising either from large-scale separative drainage systems or other local drainage systems encompassing different types of measures such as green and blue roofs, green walls, bioswales and rain gardens, among others. Collected rainwater is then directed to sunken areas within the public space, which store the water and tolerate its temporary presence. Besides serving as open-air reservoirs, these sunken areas can be designed to encompass different uses in the dry periods, such as amphitheaters, playgrounds and fairs, among others. By making water visible, people can be closer to natural water cycle changes and thus reduce their vulnerability toward those processes, a feature that is particularly valuable in adaptation endeavors.

According to Rotterdam's second "Waterplan," water plazas can be divided into the following dissimilar subtypes: "the water balloon," "the deep square," "the shallow square," "the dam," "the recessed square," and the "smart street profile" (Sliedrecht et al., 2007, p. 98), all of which have the distinguishable characteristic of being a particularly suitable adaptation measure to be applied in inner-city areas (Boer et al., 2010), especially where outflows are hindered, such as in situations when high groundwater levels impede infiltration. Another feature particular to water plazas is that their design is inherently associated with the design of urban public spaces of multiple functions. Resulting designs therefore entail variable dimensions, and thus different water retention capacities, as well as encompassing vegetation or solely inert materials.

Lastly, it is important to remember the maintenance requirements implied when applying this measure. Bearing in mind that the water that reaches the sunken areas of the square is unlikely to have been previously purified, it carries with it some pollutants and solid materials such as mud, litter, leaves, and branches. This waste remains on the square after water volumes have been redirected. As such, these solid materials need to be removed when the square becomes dry after a storm so that the designed urban area can maintain its attractiveness for other public space uses.

4.3   Underground reservoirs

Underground reservoirs, also known as regulation reservoirs, are a highly specialized drainage infrastructure measure. They are generally large in scale and therefore difficult to implement in compact urban areas.

At first, it can be difficult to conceive how this measure relates to public space. However, as the city of Barcelona has taught us, with its more than 11 implemented reservoirs, it is possible to have this large-scale infrastructure integrated with the design of public space. Albeit with some limitations and requirements, Barcelona's public spaces existing above underground regulation reservoirs range from parks to sport fields, from squares to streets.

Of all the types of measures comprised within the category of reservoirs, underground regulation reservoirs are the only measure that may receive more than just rainwater. More specifically, this type of measure is usually integrated within preexisting combined sewage systems that mix both sewage and stormwater. One of the main goals of this measure is therefore to avoid combined sewage overflows (CSOs). Once stored, the stormwater volumes combined with sewerage flows can be gradually sent to the wastewater treatment plant (WWTP).

Underground regulation reservoirs can store very significant volumes of water. Barcelona's reservoir at Zona Universitària, for instance, stores up to 105,000 cubic meters of water, equivalent to about 42 Olympic swimming pools. Bearing in mind its size and necessary structuring system as well as the deep excavation requirements, the cost associated with implementing this measure is also considerable.

As highlighted in Matos Silva (2011), among the mandatory equipment that has a direct interference in the design of public spaces are ventilation systems with a common height of 2 m, expansion joints, hydrants, emergency exits and technical galleries. In addition, plantations are limited to the capacity of the reservoir's cover slab. Trees, for instance, are generally installed over the reservoir's pillars. Bearing in mind these prerequisites, "it's up to the designer to choose whether to use them in a discreet or manifest way" (Matos Silva, 2011, p. 39).

The case of Joan Miró park is an example of how the aforementioned requirements can be harmonized in the design of a public space. One of the highlighting characteristics of this design is an extended soil bed, raised 0.3 to 0.8 m from the ground level, which forms vegetated land ramps of slight inclination. In this case it was further possible to combine a multi-storey subterranean car park in-between the reservoir and the public space, which has likely diminished the costs of such great infrastructural investment.

4.4 Cisterns

Cisterns are an ancient concept consisting on an impermeable container for storing liquids, generally harvested rainwater that can serve for later use. They can be built either below or above ground, or even on a roof. Today, some ancient cisterns have turned into public monuments, such as the case of the Basilica Cistern (Yerebatan Saray) in Istanbul, Turkey. However, in the scope of this analysis, cisterns are understood as reservoirs of relatively small dimensions that store rainwater runoff from nearby areas. These storage chambers can be built with various materials in accordance with their specific implementation contexts, from concrete to plastic. They can be distinguished from underground regulation reservoirs for their dimensions and water basin coverage, as well as their specification to only receive rainwater. While detention basins and underground reservoirs are usually constructed to collect runoff from a wide area such as a housing estate or industrial area, cisterns are here understood as reservoirs that collect runoff from nearby adjacent areas of a block, namely streets, gardens, and planters, among others. Cisterns can also exist right below pavements, a recent technique that is often mentioned in the literature as "reservoir pavements."

The landscape architecture project (by Olin Partnership) of the Ray and Maria Stata Center at MIT in Massachusetts, US, is a good example of an autonomous water management system integrated in the design of a public space. Within this system, composed of several measures, there is a cistern below a constructed wetland. This cistern is composed of stacked, injection-molded plastic panels, and in between the cistern and the wetland there is an isolating geosynthetic clay liner (Zheng, 2007, p. 47). The cistern collects and stores rainwater that is previously filtered by passing through soil and hydrophilic plantings. Collected water in the cistern is used for irrigation of the gardens and toilet flushing.

For an example of the so-called reservoir pavement, i.e., a cistern right below a pavement structure, one can consider the case of Georgia Street in Indianapolis, also in the US. In a subtle yet significant design, this street collects rainwater runoff into its central permeable boardwalk, below which there is a water management system that alleviates the city's overloaded combined sewers. Collected rainwater from the street is first directed to a stormwater forebay that serves for irrigation. When forebay troughs below the permeable pavement are full, overflows run off into an infiltration basin composed by a sandy subsoil, which will further filter water before recharging the groundwater aquifer. According to Ratio Architects, this system will reduce Georgia Street's runoff by more than 50 percent during a 10-year rain event and 40 percent during a 100-year storm.

5 Bioretention

Bioretention measures can be generally characterized by their capacity to simultaneously include the processes of retention, infiltration and evapotranspiration. They consist on excavated landscaped areas in which runoff is collected and infiltrated into the soil and, when necessary, into an underdrain connected to the main drainage network. Measures included in this category generally entail a broken stone bottom layer and engineered soil in order to offer high infiltration rates and efficient pollutant removal as well as fertile growing conditions for vegetation. Planted vegetation should preferably be riparian, from herbaceous species to woody bushes and even trees. In addition, overflow mechanisms, such as surface dischargers, should be considered when overspills to adjacent areas are unwanted.

The implementation of these measures, particularly when associated with planted vegetation, may significantly reduce flood occurrences. Among the "useful results" of bioretention measures, Richard Forman highlighted the following:

> more friction, more infiltration, more subsurface water flow, more groundwater recharge, a higher water-table, less subsidence of the surface, more evapo-transpiration, less surface-water flow, less erosion, less sedimentation, longer lag time to peak flow, lower peak flows, and less flooding.
>
> (Forman, 2014, p. 173)

Other benefits include climate melioration, namely through the reduction of albedo ratios; environmental improvement, namely by bettering water and air quality; consequent natural biodiversity increment; as well as aesthetical enrichment.

Among the measures encompassed in this category are "wet bioretention basins," "dry bioretention basins," "bioswales," "bioretention planters," and "rain gardens."

None of these measures is recommended for areas with contaminated soils, as the infiltration of polluted water is not intended. In these situations, detention ponds (such as the "artificial detention basins" mentioned earlier) are a more suitable measure.

Some important implementation and maintenance specificities are furthermore common to all the aforementioned types of measures. Among them are the requirements of any landscape design, such as the short-term maintenance in the first and second years. Other maintenance needs include the removal of trash and debris, the monitoring of clogging in the drainage and overflow mechanisms and the removal or trimming of undesired plants. The design must further withstand natural and human disturbances such as droughts or severe rain, as well as vandalism or chemical spills. Altogether, a varied range of planted vegetation should be favored over monocultures. In addition, one must recall that as vegetation matures, the bioretention system becomes more efficient and less costly to maintain.

5.1 *Wet bioretention basins*

Wet or dry bioretention basins aim to reduce overall stormwater runoff, minimizing the outflow volumes during extreme climatic events and alleviating the pressure upon existing drainage systems. These measures additionally promote the infiltration of retained water volumes through a built bottom substratum and surrounding slopes, favoring aquifer recharge. These basins can be either "wet" or "dry," meaning that they can either have a permanent water level or only temporarily hold stormwater during extreme events.

These measures are compatible with recreation activities, especially when including vegetated areas that further contribute toward environmental comfort and ecological quality. Both wet and dry bioretention basins can be designed in a way that promotes public use, contrasting with conventional engineered retention basins, which often comprise basins with clear boundaries for public use. Collected and stored runoff under these measures may additionally serve as water supply for irrigation and firefighting purposes.

Specific maintenance procedures associated to these measures include water quality control of the stormwater inflow; water quality control of stored water; removal of floating elements; cleaning of drainage devices, specifically by unclogging storm outlets or grid chambers; and protection, treatment and cleaning of the bottom and edges of the basin (LNEC, 1983, p. 256).

In the case of wet bioretention basins, additional periodic stormwater volumes are stored above a permanent water level. The depth of the permeant water level should range from 1 to 3 meters, with it being particularly important for the marginal slope to be gentle in its gradient so that the risk of drowning is minimized (Novotny *et al.*, 2010, p. 205).

Among the many examples of wet bioretention basins is the Qunli Stormwater Wetland Park designed by Turenscape. In this notable case, together with an ecological restoration of a preexisting wetland, a new public park was designed. Among other features, this park offers an intricate network of paths, some at a groundwater and others at an elevated level, from which the ecological processes of the landscape, particularly those related to the water cycle, can be appreciated.

5.2   Dry bioretention basins

Dry bioretention basins are also very common in mainstream drainage endeavors. During dry weather, and depending on the designed scale, they may welcome other uses such as temporary markets or football fields. They are here distinguished from the submergible park type of measure, generally because of their form and common location. While the submergible park type of measure is generally located in downstream areas adjacent to a stream that is particularly vulnerable to fluvial floods, dry bioretention basins are generally located in upstream areas and usually serve the purpose of buffering pluvial floods. On the other hand, while the submergible park type of measure is generally shaped with gentle slopes toward the stream, dry bioretention basins are usually shaped in a concave form. Dry bioretention basins may also resemble rain gardens when designed on a small scale, yet they usually encompass less vibrant and specific vegetation and rather include a highly porous substratum for stormwater to be filtrated and infiltrated.

Dry bioretention basins must not be implemented in areas with high groundwater levels, as this may lead to prolonged standing water and consequent intrusion of untreated contaminants into the soil, as well as mosquito breeding. In the same line of reasoning, it is a measure unadvised for adaptation to groundwater floods.

Parque da Cidade in Porto, which will be further described in the next chapter, is an example where this measure is applied in various forms. In this example, one may note the inclusion of sustaining walls within the limit of lower topographic elevation that may also be used as benches, reinforcing the possibility of multiple public space uses.

5.3   Bioswales

Bioswales, bioretention planters and rain gardens entail numerous common characterizing aspects, one of the most important being the incorporation of phytodepuration systems of vegetation.

Similar to dry bioretention basins, these measures must avoid high groundwater levels. In addition, their constructed underground layers, composed by engineered soil and open-graded stone, should be sufficiently deep in order to properly filtrate and treat stormwater. These measures are also particularly inappropriate for areas with highly contaminated runoff, given the limited purification capability of phytodepuration systems of vegetation. They are also not adequate for drainage areas with great storm volumes and velocities due to the consequences of erosion, which may significantly reduce their infrastructural capacity to store runoff and thus attenuate peak flows.

Bioswales, bioretention planters and rain gardens have been commonly used to provide traffic calming and delineate parking bays. It is, however, important to note that these three measures should be located some meters away from buildings in order to prevent foundations from being damaged by moisture. Lastly, when designed in relatively small areas, these measures may further endorse community involvement. Some examples have shown how residents have namely shared the responsibility of maintaining such ecosystem-based flood adaptation measures, such as the case of the bioretention planters of the Derbyshire Street Pocket Park (Figure 2.5) in or the rain garden at the heart of Bridget Joyce Square, both in London, UK.

Regarding bioswales, they can be generally characterized as a linear vegetated ground channel designed to collect, treat, infiltrate and, specifically, convey runoff. In addition, stormwater drained from the streets and sidewalks may this way flow directly toward the

Figure 2.5 Derbyshire Street Pocket Park, London, UK.
Image credits: Author's personal archive, 2019.

roots of street trees (Forman, 2014). Bioswales are generally shallow, have a variable section and are commonly covered by native grasses and/or small bushes. This measure therefore also reduces runoff and is capable of naturally filtering some pollutants through the vegetation and underlying engineered underground layers. Bioswales may be easily retrofitted into existing urban areas, especially along streets or integrated in the design of parking lots and other open space areas of public usage. Besides their infrastructural value for flood reduction and biodiversity enhancement, bioswales can additionally contribute as physical boundaries or simply for the encompassing aesthetic comfort value.

Among the several existing examples of implemented bioswales, the case of the High Point redevelopment in Seattle, US, illustrates well how this measure can be integrated in the design of its public spaces.

5.4 Bioretention planters

Bioretention planters can be generally characterized as low vegetated flowerbeds, sequentially implemented along a road, commonly composed of various scrub and small tree species bounded by a curb. In some examples, bioretention planters are also fenced along their limits. These specialized planters include a gravel and rock medium that filters collected stormwater and promotes its infiltration into the soil (Figure 2.6 and Figure 2.7).

Unlike the bioswales, which are generally targeted at the transport and depuration of stormwater, bioretention planters are mainly aimed to store and filtrate water within each "planter" and to gradually promote infiltration into the subsoil. This measure is therefore most efficient when associated with other measures that capture and convey stormwater, such as green roofs or bioswales.

Figure 2.6 Design proposal of a bioretention planter for Avenida Gago Coutinho in Lisbon, Portugal. Developed for a thesis in landscape architecture at the University of Lisbon, Portugal.

Source: José Barradas, 2019.

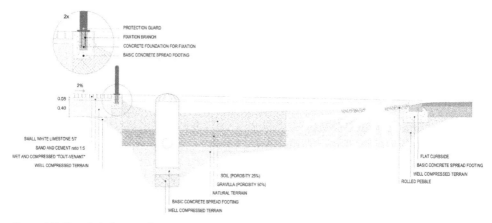

Figure 2.7 Detailed design of the bioretention planter proposed for Avenida Gago Coutinho in Lisbon, Portugal. Developed for a master's degree dissertation in landscape architecture at the University of Lisbon, Portugal.

Source: José Barradas, 2019.

These specially designed landscaped planters may fit within sidewalks, road separators and park boundaries, among others. Although their design may vary in shape, they usually have rectangular edges. In turn, their edges can include additional functions besides limiting and containing stormwater, such as benches, bicycle racks and fences, among others.

This measure is commonly built between existing surface structures and elements such as driveways, signs, street furnishings and street trees. In some cases, it serves to separate pedestrians from the moving traffic. In this latter situation, a suitable sidewalk that accommodates both the bioretention planter and pedestrian circulation is required.

In order to better exploit the capabilities of this measure, the existing soil should not be too porous, as it may reduce the treatment capacity and increase the risk of groundwater contamination. Likewise, reduced permeability may lead to the clogging of the system. In keeping with this method, the use of sedimentary traps is advisable. For an illustrative example of a bioretention planter, one may consider the project at Bakery Square in Pittsburgh, US.

5.5 Rain gardens

Rain gardens, still within the category of bioretention, are specifically envisioned to imitate the natural rainwater absorption of a forest or meadow. They consist of low, concave, areas planted with a variety of deep-rooted filtrating shrubs, perennials, and trees that receive and infiltrate rainwater (Figure 2.8 and Figure 2.9). The commonly varied and intense use of vegetation species further promotes a rich localized biodiversity.

The design of rain gardens is very diverse, albeit most are irregular in form. They are usually bigger than bioretention planters and smaller than retention and infiltration basins. Rain gardens can be incorporated within most outdoor urban landscapes, although they are more commonly found in residential yards, squares, and parks.

Taasinge Square in Copenhagen, Denmark, integrated in the municipal Climate Adaptation Plan, is one example of a rain garden among many others, usually of smaller scale, implemented within urban developments.

Figure 2.8 Design proposal for a rain garden at the intersection between Avenida Gago Coutinho and Avenida D. Rodrigo da Cunha in Lisbon, Portugal. Developed for a master's degree dissertation in landscape architecture at the University of Lisbon, Portugal.

Source: José Barradas, 2019.

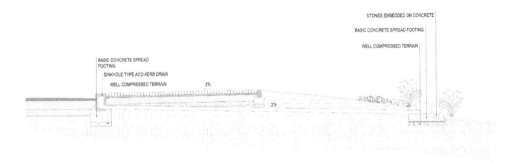

Figure 2.9 Detailed design of the rain garden proposed for the intersection between Avenida Gago Coutinho and Avenida D. Rodrigo da Cunha in Lisbon, Portugal. Developed for a master's degree dissertation in landscape architecture at the University of Lisbon, Portugal. Permeable pavement.

Source: José Barradas, 2019.

Permeable pavement is a pervious surface that allows water to infiltrate through its surface and into underlying engineered layers. From the storage layer, water can infiltrate into the ground, or it can be reused or conveyed into other measures such as a raingarden or a cistern, for instance. Permeable pavements can therefore attenuate runoff and can remove common street pollutants such as hydrocarbons and metals through the absorption and filtration processes occurring in their pervious surface and hidden sublayers.

Measures associated with this category are commonly used for source control purposes, as they can essentially manage rainfall that directly lands on their surface. They are generally able to cope with severe storms and are beneficial for adaptation to pluvial, fluvial and coastal floods. However, some potential constraints are important to bear in mind, such as the need for the soil to have high infiltration rates, the need to avoid soils that can be easily saturated from a high water table and the need to acknowledge the robustness of the pavement in accordance with the expected weight and volume of traffic (Philip, 2011, p. 35), among others.

Indeed, a permeable pavement can become impermeable if the aforementioned factors are not taken into consideration. In other situations, a paved surface may reach its maximum load capacity as a consequence of being frequently used and/or by sustaining heavy weights, compressing the pavement's structuring layers to a point where no water infiltrates. This phenomenon is frequently observed in densely populated and compact urban areas where the typical Portuguese pavement *Calçada Lisboeta* is applied. Despite it being a pavement that is generally characterized by being semi-permeable, in most situations, namely in Lisbon's center, this pavement is entirely impermeable.

Among the highlighted measures within this category are "open cell pavement," "interlocking pavement" and "porous pavement." Open cell pavement allows water to enter through the gaps existing within the unit of pavement itself; interlocking pavement allows water to enter in between the gaps of pavement units; and porous pavement is made of aggregate and porous materials that enable water to pass through its surface. All of these types can be designed for car or pedestrian traffic and can be implemented in almost any urban area. Presently, there is a wide variety of pavement designs and materials, which can further add to the aesthetic quality of a space. Their texture can also serve to distinguish areas or to manage traffic velocities.

5.6   Open cell pavement

Open cell pavement is made of pavement units that encompass a void in their form, therefore paving surfaces that incorporate gaps of materiality. Generally, these pavement units are prefabricated, yet the design can generate various forms of voids. The materiality of this type of pavement can range from concrete to plastic.

The parking lot adjacent to the dock of Alcântara in Lisbon, Portugal, is one of the many existing examples of open cell pavements (Figure 2.10).

5.7   Interlocking pavement

Interlocking pavement is here considered as a pavement of interlocking blocks (Figure 2.10). These blocks can be of regular or irregular form. Their materiality may also vary, from different types of stone to different types of concrete and aggregates. The distance applied between blocks may vary; regardless, it must not be too wide so that they no longer serve the function of pavement, nor too narrow, so that they no longer include permeable interstices.

Open cell pavement and interlocking pavement are permeable pavements that may include gravel, soil or grass in their interstices. In both situations, stormwater may infiltrate downward and evaporate upward.

5.8 *Porous pavement*

Porous pavements are made out materials that are full of voids in their composition (Scholz and Grabowlecki, 2007). These voids allow some of the stormwater to infiltrate into the apertures and downward into the sandy soil below (Forman, 2014). These materials can range from pervious concrete to porous asphalt or a simple aggregate of grit or broken stone. As technology evolves, specialized porous pavement's increase their absorbing capacity. For example, Lafarge Tarmac, a British building materials company, has come up with a new type of concrete in recent years that has the capacity to absorb 4000 liters of water in around a minute. Regardless, it is important to note that, over time, these pavements are expected to clog and therefore reduce their infiltration effectiveness. Expensive procedures may be used to vacuum or pressure-wash these clogs. As such, today, this type of measure is still more suitable for areas of reduced load capacity such as sidewalks or park paths.

Despite being substantially different in structure, porous pavements can be visually similar to nonporous pavements. The work at Percy Street in Philadelphia, US, can serve as an example of porous asphalt applied in a small urban street. Another example of a porous pavement is grit (thick sand with small stones to blend), which is frequently seen in historic gardens (Figure 2.10).

Figure 2.10 From left to right: Open cell pavement at a parking lot in Cascais, Portugal; interlocking pavement of a sidewalk adjacent to Central Park, New York, US; saibro/grit pavement at S. Miguel Pateo, Évora, Portugal.

Source: Author's personal archive, 29 June 2019, 1 April 2019 and 22 May 2019, respectively.

6 Infiltration techniques

In the scope of this analysis, the types of measures associated to this category are primarily targeted at fulfilling the purpose of stormwater infiltration. In doing so, these measures contribute not only to the reduction of surface runoff but also the aquifer recharge. This approach does not neglect the fact that other types of measures, integrated within other categories, might also encompass the same infrastructural function, albeit as a secondary purpose.

Among the measures specifically targeted at stormwater infiltration, one may consider infiltration trenches, leaky wells, geocellular systems and green gutters. Yet only "infiltration trenches" and "green gutters" were substantiated with implemented examples that considered these measures as constituent elements of a public space design.

The types of measures within this specific category further promote the efficient removal of suspended solids, organic matter, soluble metals, and nutrients. The construction of these measures features underground layers that may be composed of gravel, sand or broken stones. The design criteria, such as the materials to use in the sections of each underground layer, are highly dependent on rainwater intensities, the condition of local soil and the available space. They can be implemented in various urban settings from public gardens and parks to roadside alignments, from parking lots to roundabouts.

Overall, these measures provide a significant reduction of runoff volumes while further contributing to the treatment of stormwater through the removal of transported sediments and light pollutants. Yet measures within this category are not appropriate for areas that receive runoff with high pollution levels, such as train oils or chemical spills. They are also inappropriate and useless when subsurface soil is not sufficiently permeable or for areas close to the phreatic level.

For prevention purposes, raised drains should be provided when implementing measures within this category. If the infiltration systems, trenches, or green gutters reach their retention capacity after an intense storm, stormwater should be redirected to where it is most convenient. This same concern is applied to the types of measures within the category of bioretention.

6.1   Infiltration trenches

Infiltration trenches collect and retain stormwater within their underground layers of high porosity levels, until water infiltrates into the subsoil or evaporates into the atmosphere. This type of measure not only reduces stormwater peak volumes, which frequently compromise downstream areas but also contributes to the quality improvement of stormwater discharges through its porous filtrating layers.

Infiltration trenches can generally be characterized as having a relatively shallow depth (no more than 1 meter), a gentle concave depression and a longitudinal development. They are relatively easy to implement and are not very costly (LNEC, 1983). Constituent layers of this type of measure generally include a gravel layer, where sediment, leaves and debris are trapped, surrounded by a geotextile fabric.

6.2   Green gutter

A green gutter is a thin and shallow landscaped strip that can be located between a road and its sidewalk. In similarity with infiltration trenches, this type of measure is developed longitudinally. It is also designed to manage stormwater runoff, mostly through the process of infiltration, yet with inferior volume capacity. In accordance, this measure is commonly placed below the street's gutter, and it is particularly appropriate when there is a reduced area for implementation. Other benefits include the improvement of the overall amenity of the streetscape and the provision of a physical separation between lanes with uses of different velocities (such as pedestrian and traffic lanes). The edge design of green gutters should bear in mind the inhibition of pedestrians, cyclists or cars from falling into the strip. It is therefore a measure that may not be considered suitable for areas of intensive pedestrian use, such as exterior markets, children's playgrounds or busy park areas. Its management requires the desirable periodicity of an adequate green area maintenance.

7   Stream recovery

Stream recovery, here only focused upon urban territories, essentially involves the processes of improving or recuperating, totally or partially, the natural ecosystem of a watercourse. This process may occur in smaller or more extensive sections of a stream. It is, however, important to note that stream recovery is only possible when major pollution sources have been eliminated, especially arising from CSOs or SSOs (combined sewage overflows or sanitary sewage overflows).

In dense urban settings, the ecological potential of a stream recovery adaptation project can rarely be comparable to that achieved in a less urban setting. Regardless, recovered streams must be ecologically viable and sound, as they will technically become part of the surface drainage system. Other benefits are more likely enhanced in urban settings, specifically the ones that respond more directly to human uses and necessities, such as recreation and enjoyment and a closer connection to overall ecological processes.

Public spaces generated out of stream recovery projects are increasingly common. Considering the extended literature on the subject of urban rivers and their recovery, specifically research works from Graça Saraiva or Mathias Kondolf, or feasibility studies applied to particular contexts (such as Lehrer et al., 2010), a significant number of examples, as emphasized in this section, can be considered as references in the improvement of a community's quality of life. These measures, progressively considered as flood adaptation undertakings, are further applied for most types of floods (pluvial, fluvial, groundwater, artificial drainage, coastal). Measures within this category additionally contribute to the growth of the river's water flow capacity, thus further reducing the velocity of water discharge during storm events and its associated peaks of erosion. By improving the riparian habitat of streams, these measures also harvest stormwater, promote groundwater recharge and enrich biodiversity.

Among the identified processes of stream recovery in the scope of this analysis are "stream rehabilitation," "stream restoration," and the process of "daylighting streams" that were formerly concealed.

7.1   Stream rehabilitation

Stream rehabilitation is here understood as a partial improvement of a severely disturbed open stream. A severely disturbed stream includes, for instance, a natural stream that has been channelled by an open-air half-piped concrete conduit. Stream rehabilitation is therefore essentially targeted at resurrecting the stream back into an improved working order of adequate ecological quality. When applying this adaptation measure, the visual aspect of that artificialized stream benefits will improve together with its ecological status.

An example considered as stream rehabilitation is the intervention made in Ribeira das Jardas at Agualva-Cacém, Sintra, Portugal, financed by the Polis Program (2001–2008) and designed by NPK landscape architects. This project gave rise to a new urban park by a watercourse that was previously limited by a concrete wall.

7.2 Stream restoration

Stream restoration is here interpreted as the change from an artificial stream to a "near-natural" stream. It can be very difficult to recreate the watercourse ecosystem exactly like it was before its disturbance period, simply because of the required change of its surrounding situation, including all contemporary urban stresses such as pollution or closing of ventilation corridors. Regardless, in a stream restoration process, the structure, function and self-sustaining dynamics of its riparian habitat is reestablished.

Among the existing examples of stream restoration is the intervention made at Kallang river in Singapore, which gave rise to the Bishan Park. In this project, designed by Atelier Dreiseitl, a 2.7 km long straight concrete drainage channel was replaced by a meandering natural river 3 km in length. Alongside the stream, 62 ha of park area were designed to accommodate the dynamic process of the restored river system (encompassing temporarily flooded areas) as well as to provide recreation and amenity value.

7.3 Daylighting streams

The type of adaptation measure here named daylighting streams essentially involves the process of bringing a buried stream to the surface. This measure reestablishes a watercourse to its former channel profile or in a new channel profile with a more or less rigid formality that can exist built between buildings, streets, parking lots or playing fields.

The number of examples that illustrate how this measure may be implemented increase every year. Among them, one specific example is the daylighting of Thornton creek in Seattle, US. This project, designed and engineered by SvR Design Company, turned an abandoned parking lot into a water treatment facility that is also an open public space to be enjoyed by the local community.

8 Open drainage systems

Open drainage systems are uncovered water channels that are complementary or alternative to underground drainage systems. These open channels should only receive water that has been previously treated or is already free from pollutants. While measures within this category include improvements made in preexisting watercourses, others serve to convey cleaned stormwater arising from other types of measures, such as rooftops or bioswales, leading it into the underground sewage system, to a receiving waterbody or to other types of adaptation measures such as rain gardens, bioretention basins or cisterns.

Among the measures presented within this category are "street channels," "extended channels," "enlarged canals," and "check dams," all of which can be designed in various dimensions and forms and all of which can be particularly effective when tackling pluvial and fluvial floods.

In resemblance to other measures that instigate a community's approximation to the processes of natural water systems, the aforementioned measures can also greatly change the perception of an urban space and its social appropriation by exposing a fragment of the water cycle and by inviting users to be a part of it, whereby education and awareness is promoted.

The goal of bringing people closer to these exposed stormwater conveyance systems can be approached in many ways, from the inclusion of bridges to the use of stepping stones. This latter concept, which was early on used by the Romans in the crossings of their streets,

Figure 2.11 Street channels type at the city of Granada, Spain.
Source: Author's personal archive, 9 June 2009.

which served as stormwater runoff channels, is now often used in contemporary designs such as in the case of the Roombeek Street water channel in Enschede, the Netherlands. Open drainage systems can be designed to also include filters, cascades, pools and many other water features that promote different possible interactions between rainwater, community and urban design.

8.1   Street channels

As the name indicates, street channels are a means through which cleansed water is transported from one point to another. These water channels can have various dimensions and their design can have many different forms, from straight canals of 0.15 × 0.15 meters sections to light street adaptations also commonly called "smart street profile" (Sliedrecht *et al.*, 2007). Among the other different possible applications is the idea to convert train or tram tracks into street channels. The aesthetic possibilities when using water channels are wide, as the city of Granada in Spain so evidently reveals (Figure 2.11). Of all design possibilities, it is important to bear in mind the requirement for the channel to have a minimum sloping gradient in order to avoid stagnant water.

This measure can be applied in dispersed urban areas as well as in compacted urban centers or coastal areas. As illustrated by the medieval example of central Freiburg Bächle, Germany, as well as by the recent interventions at the old city center of Banyoles, Spain, as highlighted in the next chapter, this measure can be further compatible within a historical center.

8.2   Extended channels

Extended channels are here considered as the artificialized water courses that have been extended in their length, either through an orthogonal or meandering design. By extending preexisting water channels through the design of new water lines, wider stormwater volumes can be managed and thus alleviate stormflow velocities, erosion and overall flood risk.

In order to illustrate an example of an extended channel measure applied in the design of a public space, one may consider the intervention made in Pier Head in the city of Liverpool, UK. The goal of this project was to create a canal extension – linking the Leeds and

Liverpool canal to the north with the dockland water basins adjacent to King's Waterfront to the south – which further included the design of a public square worthy of such a high-profile site as the surroundings of Three Graces waterfront buildings.

8.3 Enlarged channels

The type of adaptation measure here named as enlarged canals encompasses the process of broadening the basin of an artificial or natural water course. Among the existing examples, it is possible to highlight the case of the London 2012 Olympic Park. In this case, the channel of the river Lea was widened specifically for the implementation of a generous area for wetlands.

8.4 Check dams

The adaptation measure identified here as check dams can be generally characterized by the implementation of small permanent or temporary barriers along a water course in order to promote localized water accumulation and overall attenuation of runoff velocity (Forman, 2014). These barriers can be built from various materials, such as wood or stone, plastic or concrete and steel or plexiglass. This type of measure is usually associated with other types of measures that include the infrastructural strategy of conveying, that is, which include the process of transporting stormwater through channels.

This measure also reduces erosion and promotes sedimentation. After stormflows, water is retained behind the small dams, where it seeps slowly to lower soil layers or evaporates. In order to promote water purification and avoid mosquito breeding, the implementation of this measure should be accompanied by the plantation of appropriate vegetation. Moreover, this measure requires periodic maintenance and sediment collection in the upstream area of dams, preferably after each storm.

One illustrative example of applied check dams is the sloping bioswale at the Kronsberg hill residential area in Hanover, Germany.

9 Floating structures

Floating structures are an old concept and practice that has been revisited in recent years. One can mention, among several other examples, the biblical reference to the Noah's Ark; the ingenious Floating Gardens of the ancient Aztecs; the floating islands of a local tribe by the Uros in Peru and Bolivia, built with native vegetation; or the Teatro del Mondo architectonic gesture by Aldo Rossi, a floating building designed for the 1979 Venice Biennale.

Likely as a result of climate change threats, newly incorporated designs and experimental approaches have been exploring the potential of floating structures as the ultimate flood resilient measure, one that tolerates and adapts to any type of flood, being it pluvial, fluvial, groundwater, artificial drainage, or coastal flood.

Floating structures can range from floating mega structures such as ocean oil platforms, to floating buildings and floating urban developments such as the ones in Ohé en Laak, Limburg, in the Netherlands, designed by Dura Vermeer. Similarly, floating structures can also be built to support public spaces. Architecture at the among the types of floating structures that can be specifically associated to public space design, the following are high-lighted here: "floating pathway," "floating platform," and "floating islands."

9.1   *Floating pathway*

A floating pathway or a floating bridge connects one place to another through water, usually by means of a flexible structure detached from any ground supporting element. In similarity to other floating structures, a floating pathway adapts to different water levels without compromising its public space character, namely the function of being a passage.

Among the existing illustrative examples of a floating pathway, one may distinguish the design of the Ravelijn Floating Bridge at Bergen op Zoom in the Netherlands, for its simple lines and aesthetic appeal. This pedestrian bridge had two primary goals: the first was to connect a city fortress to its city center, and the second was to provide an escape route from the fortress in case of emergencies. The shape of the bridge pathway is convex, blending with the water and its surroundings through the creation of a mirrored effect. Furthermore, the stairs that connect to the small pier in the fortress entry are designed to move up and down with the water level, with fewer or more steps. In order to allow the bridge to float, air-filled polyethylene pipes are located underneath its timber surface.

9.2   *Floating platform*

The floating platform as a type of adaptation measure is here interpreted as an extensive structure contiguous to a non-floating structure which offers a public space that may encompass multiple functions and uses.

Floating platforms can be used for the design of floating gardens, such as the case of the Yongning River Park at Taizhou, China. They can also be used for the design of floating waterfronts, as is the case of the floating piers of the Landungsbrücken by the Elba river in Hamburg, Germany. Deprived from the load of architectonical reference, this former example offers a 700 m long public space that supports a diverse and varied use, from commercial to cultural, whilst at the same time providing a close relation with the Elbe river.

9.3   *Floating island*

The adaptation measure here named as a floating island is different from the aforementioned floating measures, as it refers to an occasional floating structure and not a linear path or extensive area. It can comprise an esplanade, a sports field, a stage or an audience structure for cultural events.

Floating islands can be constructed of various materials. In some cases, obsolete ships are reconfigured from their transport function into another function, such as a public space. That is namely the case of the Bathing Ship (Badeschiff) at the Spree in Berlin, Germany. In this example, an old barge was adapted into a pool. Its superstructures were removed, allowing for its hull to be flexibly moored in a jetty. With a depth of around 2 m and a length of 32 m, a floating pool by the Spree can now be enjoyed. The same ship could have been adapted to other types of public uses.

10   **Wet-proof**

The types of adaptation measures within the identified wet-proof category include different types of public spaces that are resistant to the periodic and temporary submersion by floods. Among the alternatives, "submergible parks" and "submergible pathways" were singled out. In most situations, these measures are applied in flood-prone areas, which can exist in interior lowlands or in areas adjacent to natural water streams. Wet-proof measures are particularly suitable in the adaptation to pluvial, fluvial and coastal floods.

Urban elements used in the design of these measures, such as pavement or urban furniture, must be particularly resistant in order to sustain the impacts of recurrent flood events. More specifically, they must be made out of robust materials, and particular attention must be given to the construction of solid foundations. In the same line of reasoning, the vegetation used in these elements must also be able to sustain periods of immersion, such as species originating from the habitat of riparian woodlands that can tolerate the oscillations between flood and drought. Once the risk of flooding is high in these wet-proof areas, the design of the aforementioned measures must also encompass clear and visible information and alert signage indicating waterflow forecasts and escape routes.

10.1 Submergible parks

Submergible parks as an adaptation measure are generally associated with other measures, such as the ones within the category of stream recovery. The resulting public spaces are, however, more extensive, generally configured as parks. These parks can serve a wide range of purposes such as playgrounds or sports facilities, although only during dry weather.

One may further recognize similarities between the dry bioretention basins measure and the submergible parks measure: while the first is more appropriate for upstream areas, the former is necessarily near a waterbody (fluvial or coastal). On the other hand, while dry bioretention basins are typically shaped in a concave form, submergible parks generally encompass gentle slopes directed toward the waterbody. Furthermore, dry bioretention basins are particularly suitable for pluvial floods, while submergible parks are most relevant for coastal or fluvial floors.

Among the various existing examples where this adaptation measure has been applied, the case of Buffalo Bayou Park in Houston, Texas, in the US is here highlighted. Designed by SWA Group, this project turned a flood-prone green field, which included a poorly managed stream, into an open public park that is also a stormwater management infrastructure. Among other processes, a stream restoration process included the redesign of the stream's profile, expanding it into a submergible park. As a result, when the periodic floods occur, the park can sustain the impact without considerable risk.

Information signage that alerts visitors about to enter flood-prone areas is also common within these parks. For instance, the Rio Besòs Park, further explored in the next chapter, includes such signage in the form of electronic placards near some of the park's entries, which very clearly inform whether the park is open or closed due to expected flooding.

10.2 Submergible pathways

Submergible pathways are a type of wet-proof structure that, during dry weather, can increase the available public space area for connection, leisure or other purposes. Such measure was applied in Quai des Gondoles at Choisy-le-Roi by the river Seine in France. Implemented using a flood resistant and robust steel structure that is fixed underground, this 4 m wide boardwalk extends over 500 m. When floods are anticipated, the entrances to this submergible boardwalk are closed with adaptable barriers (Prominski *et al.*, 2012, p. 164). One other illustrative example is Passeio Atlântico at Porto, Portugal, designed by Manuel de Solà-Morales and others. This submergible pathway, which develops between Montevideu Avenue and the Atlantic coast, encompasses both submergible boardwalks and submergible concrete pathways.

11 Raised structures

Raised structures as an adaptation category are here interpreted as the public space structures that are elevated or suspended over the maximum levels of a waterbody in order to be unaffected by flood events.

The concept to elevate a structure in order to protect against flooding is long-established. One might recall, for instance, the prehistoric stilt-houses settlements in lake Zurich in Switzerland (namely, Lacustrine Village). It is also a model that is still very much used in the present day. For instance, recently, after the destruction by Hurricane Sandy in 2012, thousands of homeowners from the most affected areas in the state of New York have applied for specific funding that would help them elevate their homes. Entire houses are therefore still being raised up through pillars, at levels 1.5 to 3 meters higher than they were before, in order not to be affected by future floods.

As might be expected, this concept is also often applied in the design of public spaces, namely in promenades, passageways, stages, esplanades and squares, among others. This approach is generally used in waterfront margins, although it is also possible to implement into interior flood-prone areas. In waterfronts, measures within this category are commonly used in densely urbanized areas where the availability of space is limited. Through overhanging balconies or elevated platforms, space is extended over the water as part of the waterfront structure.

The implementation of these measures only partially affects the watercourse profile and the flood basin and their stormwater volume capacities. The resulting public spaces are therefore unlikely to be affected by floodwaters from any source, as they are placed over the levels of projected flood dynamics. They can be used all year round, and their design generally fits with the surrounding open spaces. The designs of raised structures that protrude out of a waterbody further offer inhabitants the possibility to more closely engage with the water dynamics happening around and/or underneath. However, for safety reasons, these measures must encompass a proper fencing, and this enclosing barrier might strongly influence the visual connection with water (Prominski *et al.*, 2012).

Construction techniques used in the implementation of these measures include the use of stilts (pillars, pilotis) or cantilevered structures. Among the different possible types of raised structures adaptation measures, the following were identified in the scope of this analysis: "cantilevered pathways" and "elevated promenades."

11.1 *Cantilevered pathways*

Cantilevered pathways are structures that are raised over the water without including pillars in their construction, that is, they comprise areas suspended by a cantilevered system. Cantilevered pathways are also generally narrow in accordance with their construction technique. These structures are therefore mostly used for soft mobility (pedestrian, bicycle, roller skate and other types of mobility that infer soft pavement load capacities). This measure can be furthermore designed in various shapes and can be built with different construction materials.

Cantilevered pathways can be divided into two distinguishable types: (1) in an overhang, which is a suspended pathway that goes along, or is mostly parallel to, the margin of a watercourse and (2) in a balcony, which is a suspended terrace that is mostly configured perpendicularly to the watercourse. For example, in the daylighting of the Elster-Saale canal in Leipzig, Germany, it was necessary to substitute a former street with a suspended pathway or overhang. At the Green Park of Mondego in Coimbra, Portugal, several balconies were designed so that users could practice fishing or simply enjoy elevated viewpoints of wider perspectives.

11.2 *Elevated promenade*

An elevated promenade is here interpreted as a raised public space of large dimensions, generally associated with fluvial or coastal margins. These spaces entail more extensive areas and therefore require a stronger supporting structure such as reinforced concrete pillars. These areas may be built for pedestrian circulation as well as for automobile or public transport traffic; they may also include plantations of large shrubs or small trees.

For an illustrative example one may consider Bilbao's waterfront in Spain, which roughly extends from the Areatzako Zubia Bridge to the La Salve Bridge. With an average width of 18 m, this elevated promenade develops along the Nervión river. Its public space is composed by regularly placed small and medium-sized tree alignments as well as urban furniture such as benches and lamps. Occasionally it supports lightweight building structures such as kiosks or removable tents. Another example is the elevated promenade at the Tagus Linear Park at Póvoa de Santa Iria, Portugal, directing people toward the dynamics of the marsh on what was previously a neglected landscape (Figure 2.12).

12 Coastal barriers

In the scope of this analysis, the category of coastal barriers specifically includes the types of measures encompassed within coastal management, which aim to adapt territories to the impacts of storm events that lead to flood occurrences, especially the impacts from fluvial floods as well as storm surges and a sea level rise.

In this category of measures, the change of paradigm regarding flood risk management approaches (Matos Silva, 2016) is particularly evidenced. More specifically, adaptation measures within this category evidence the recognized need to adapt waterfronts so that they can combine both the requirement to defend land from floods and the promotion of a closer and more intimate relation between local inhabitants and the natural coastal dynamics.

In situations previously tackled through monofunctional flood protection engineering, approaches included major seawalls or other impenetrable types of infrastructure that would generally cut and divide the city from its waterfront. In this new era, various examples have shown that it is possible to maintain the high standards of robust engineering flood defense while also benefiting from the opportunities provided by interdisciplinary and multifunctional spaces of public utility. Along these opportunities is the potential

Figure 2.12 Elevated promenade at the Tagus Linear Park at Póvoa de Santa Iria, Portugal. Designed by Topiaris Landscape Architecture.

Source: Author's personal archive, 7 July 2019.

available space for public encounter. As argued in the first chapter, through public spaces the engagement in climate adaptation action can be enhanced, namely when the impacts of extreme weather can be made tangible. In carefully designed coastal barriers that also encompass public spaces, the power of water dynamics becomes meaningful for citizens and their livelihoods.

Measures highlighted in this category of coastal defense are: "multifunctional defenses," "breakwaters," and "embankments."

12.1 *Multifunctional defenses*

As the name reveals, multifunctional defenses are coastal barriers that encompass multiple functions. They are commonly applied in urban waterfronts where space is limited and flood protection is indispensable. Often, they are built within or over preexisting dikes or other types of Coastal barriers. The Netherlands frequently applies this measure in its urban coastlines; an example is the case of the project proposed by the De Urbanisten office regarding the improvement of the Dike at De Boompjes in Rotterdam.

The design of multifunctional defenses can be very diverse, yet it must contemplate intricate preexisting situations as well as the infrastructural requirements of contemporary flood defense. The possible multiple configurations of multifunctional defenses in light of different supporting contexts have been exhaustively studied in literature. Dutch research institute Deltares, for example, under the FloodProBe research project, has differentiated the following types of design concepts: cofferdam, step dike, l-wall, soil-retaining wall, oversized inner slope, and oversized outer slope (Deltares, 2013).

One example of a multifunctional flood defense is the Elbe promenade in Hamburg, designed by Zaha Hadid Architects. In this case, the previously existing coastal defense structure was outdated and had little added aesthetic value. Through this renovation project, a new and improved flood barrier was integrated with a 750 m long promenade with large and all-encompassing steps toward the waterfront. In these steps and when possible, people can get closer to the river and enjoy the dynamics of its landscape.

12.2 Breakwaters

Breakwaters or wave-breakers are structures constructed on coasts in order to block or attenuate the intensity of waves, currents or longshore drift. In most situations, these structures are used for their distinct infrastructural purpose and encompass no other function. Regardless, there are some examples that were designed with the additional goal to encompass a space for public use. When permitted by the weather and intensity of water dynamics, breakwaters can therefore also serve as viewpoints, pathways or stopovers. This is the case of the breakwaters smoothly designed for Jack Evans Boat Harbour in Australia by Aspect Studios. Other examples include the sculpturally treated breakwaters at Zona de Banys del Fòrum designed by BB+GG Arquitectes or the Barra do Douro Jetty designed by Carlos Prata Arquitecto that will be briefly explored in the next chapter.

12.3 Embankments

Embankments can be described as land reclamation constructions along a riverbank. They are mostly present in cities whose morphology is naturally elevated and thus protected against floods. Unlike dikes or levees, which are occasionally built inland, embankments are considered here as exclusive to waterfronts.

Most embankments were built during the industrial era. In similarity with the measures identified in the category of raised structures, embankments allow a growth of marginal land (i.e., by the edge of a water course) within the city.

Embankments are very common in postindustrial urban territories and, depending on the context, their "reclaimed" areas offer new land that is managed by – generally more than one – municipal stakeholders. Among the potential benefits of "extra" land is the construction of public spaces, as happens in part of Lisbon's riverfront.

13 Floodwalls

Floodwalls can be characterized as being artificial walls, generally consisting of extensive reinforced concrete vertical platforms perpendicular to the ground and alongside a watercourse. Their size and shape may vary considerably depending on the projected flood characteristics. They are mostly applied with the goal to face fluvial and coastal floods.

They are mostly used in dense urban settings where available space is very limited. In similarity to other measures that must comprise applied engineering techniques, floodwalls are not required to have an infrastructural flood defense purpose as their sole objective. Although in most situations they are implemented without aesthetic considerations, they do have the potential to be integrated as part of a public space design.

Today's know-how allows for floodwalls to be artistic sculptural elements or didactic structures while maintaining their robustness and efficiency as a flood barrier. That is the case of the adaptation measures here highlighted, namely "sculptured walls" and "glass walls."

13.1 *Sculptured walls*

The case of Main's riverside in Miltenberg, Germany, is a good example of a sculpted floodwall integrated in the design of a public space. With changing slopes and a base of varied thickness, the wall enriches the surrounding public space as an individual sculptural element. Besides encompassing a car park, green open spaces and pedestrian and bicycle paths, the public space adjoining to this sculpted floodwall is also used as an event area, namely for the Michaelismesse fair. At night, its design is exacerbated by means of illumination. During extreme flooding, the height of the wall can be further increased with additional automatic floodgates, thus also protecting the area to the one-in-100-year flood events (Prominski *et al.*, 2012).

13.2 *Glass walls*

Glass walls have been applied, for example, at Westhoven, Cologne, in Germany. They are able to provide both effective protection and unobstructed views. This flood protection means, evidenced as highly robust, further offer interesting perspectives of the water dynamics during a flood event (its rise in time, the water turbulence, the flow direction and the color of the water, among others). This type of measure can be further used for educational purposes, namely when implemented by fish pass dams such as the case of "Passage309 Fish Ladder" at Gambsheim in Alsace, France.

14 Barriers

This category of adaptation measure refers to the group of flood defenses that are locally implemented either through temporary or permanent demountable mechanisms. They are mostly applied when facing fluvial and coastal floods.

Temporary barriers can generally be described as a provisional flood protection mechanism that is composed of detachable flood protection products that are exclusively installed during a flood event and are totally removed once flood levels are no longer a nuisance (EA, 2011). This type of mechanism can be applied through the use of prefabricated or artisanal metal plaques, sandbags, inflatable devices or the combination of all the aforementioned. Prefabricated plaques are generally placed in linear series, sandbags are generally piled and artisanal metal plaques are often seen in the doorways of the more frequently affected properties.

This type mechanism only exists during a flood event. Regardless, it is a frequently implemented measure and a constituent part of the urban environment. Its assemblage process and resulting configuration should therefore be thought of, together with urban design concerns. Indeed, there is great potential to integrate and accept temporary flood defenses as established components of urban and public space design.

As the developed portfolio screening was unable to identify at least one existing example where the application of temporary barriers was integrated within the design of a public space, only the type of adaptation measure of a permanent "demountable barrier" is here considered.

14.1 Demountable barrier

A demountable barrier is here understood as a permanent flood protection, albeit movable, that has been preinstalled and requires operation during a flood event; or a structure that is partially installed in pre-implemented guides in a preconstructed foundation (EA, 2011).

Demountable barriers can include various types of mechanisms, that is, variants, from sectional barriers such as an elevated vertical wall, flood gates in the shape of a frame wall or flood doors. Several examples may serve to illustrate how this measure may be applied in practice, one of which is the Waalkade promenade at Zaltbommel, in the Netherlands. In this specific case, the height of the flood protecting wall can be further raised by 50 cm with the use of mobile mechanisms. According to Prominski, Stokmann *et al.*, a flood taskforce can raise this "wall within a wall" barrier in five hours (2012).

Other examples include the sectional barrier at Main's riverside sculpted wall in Miltenberg, Germany, or the flood gates in the Landungsbrücken building, a 1910 Hamburg fluvial station adapted to be protected from the floods of the Elbe. On a smaller scale, one may further consider the northwest gates that give access to the underground parking lot of the Gulbenkian Foundation in Lisbon, Portugal, which also function as cofferdams avoiding the entrance of (recurrent) floodwaters (Figure 2.13).

Figure 2.13 Left: Northwest access gates of the Gulbenkian Foundation in Lisbon, Portugal. Right: Detail of a northwest gate of the Gulbenkian Foundation in Lisbon, Portugal. These gates, located in the lowest topographical point of the Gulbenkian Foundation Garden, are prepared to function as cofferdams, keeping floodwaters from entering existing underground floors.

Source: Author's personal archive, 25 July 2019.

15   Levees

Levees are an old category of flood defense infrastructure. They can also be called levée, dike, embankment, floodbank, or stopbank. This category can be generally characterized by encompassing natural or artificial slopes that regulate water levels. These slopes are usually made out of earth, although they can be reinforced by strengthening their inner core with other types of materials such as steel or by improving the characteristics of their surface in order to contribute to the overall stability of the levee structure. Levees can also be reinforced by heightening or broadening their scale. They are often implemented parallel to the watercourse either in urban or agricultural lands.

In most situations, levees are unrecognized by being designed together with the design of parks or other types of public spaces. In these cases, the type of adaptation measure of 'gentle slope levees" is identified. Within this type of measure, one may further distinguish two variants: the urban levee and the green levee. Arguably, the "super-levees" could be another type of measure within the category of levees. Yet super-levees' were not considered in the scope of this analysis, as their scale, with heights over 10 meters and extensions of kilometers, is more territorial, surpassing the scale of a public space.

15.1   Gentle slope levees

The Netherlands has a great network of levees, all of which are a constituent part of spatial planning and urban design. Most of their levees have roads on their upper level, therefore serving also as transport facilities. In addition, most Dutch cities are comprised of several consecutive levels of dikes. In some situations, these dikes are so old and so strongly incorporated within the urban fabric that they can easily become unnoticed. In these cases, multiple public spaces may arise over them. That is particularly the case of the "Hilledijk" in Rotterdam. In other situations, their form can be more simply acknowledged, namely when landscaped. An example of a green levee is, for instance, the case of Corktown Common Park in Toronto, Canada, designed by Michael van Valkenburgh Associates partnered with Arup.

Flood adaptation categories and types of measures applicable in the design of public spaces

Overall, it was possible to identify and systematize 40 types of flood adaptation measures, covered by 16 categories (Table 2.1 and Figure 2.14).

In light of the disclosed results, one may note a strong correlation with the principles established by urban drainage management systems, such as SUDS, LID, BMPs, WSUD and more (Fletcher *et al.*, 2015). Yet these differences essentially rely on the initial leading focus on adaptation to the risk of flooding; more specifically, on the aim to include structural measures that not only support strategies of "prevention," such as urban drainage management measures, but that also support "protection" strategies, such as flood defense measures (European Commission, 2004). With regard to the presented list of types and categories of measures, its distinctiveness can be found in the measures associated with the "avoid" infrastructural strategy, such as breakwaters, sculptured walls, glass walls or demountable barriers. Other particularly targeted measures

Table 2.1 Flood adaptation categories and types of measures applicable in the design of public spaces

Flood adaptation measures applicable in the design of public spaces

Categories	Types	
A. Urban greenery	1. Green walls	21. Stream restoration
B. Urban furniture	2. Inverted umbrellas	22. Daylighting streams
C. Rooftop detention	3. Art installations	23. Street channels
D. Reservoirs	4. Green roofs	24. Extended channels
E. Bioretention	5. Blue roofs	25. Enlarged canals
F. Permeable pavement	6. Artificial detention basins	26. Check dams
G. Infiltration techniques	7. Water plazas	27. Floating pathway
H. Stream recovery	8. Underground reservoirs	28. Floating platform
I. Open drainage systems	9. Cisterns	29. Floating island
J. Floating structures	10. Wet bioretention basins	30. Submergible parks
K. Wet-proof	11. Dry bioretention basins	31. Submergible pathways
L. Raised structures	12. Bioswales	32. Cantilevered pathways
M. Coastal barriers	13. Bioretention planters	33. Elevated promenade
N. Floodwalls	14. Rain gardens	34. Multifunctional defenses
O. Barriers	15. Open cell pavement	35. Breakwaters
P. Levees	16. Interlocking pavement	36. Embankments
	17. Porous pavement	37. Sculptured walls
	18. Infiltration trenches	38. Glass walls
	19. Green gutter	39. Demountable barrier
	20. Stream rehabilitation	40. Gentle slope levees

may be further identified in the remaining infrastructural strategies, namely inverted umbrellas, art installations, underground reservoirs floating platforms, and elevated promenades, among others. A similar parallel can be made with encompassing concepts of greenways or green infrastructure (Ribeiro and Barão, 2006); measures included in this systematization can relate and sometimes be part of these notions, yet they are not bounded by them.

Considering the premise presented by this research, which considers the application of local adaptation measures in the design of public spaces as determinant for the quality of future cities, this chapter addresses the categories and types of flood adaptation measures identified in the proposed systematization process. Bearing in mind the fundamental requirement to be sufficiently elucidated with regard to existing knowledge and practice when approaching any public space design project with flood adaptation capacities, the identification, characterization and organization of a wide range of existing types of measures is particularly relevant for anyone involved in this type of practice. The possibility to easily grasp an overview of the existing range of options regarding the different types of adaptation measures facilitates and accelerates the initial phase of a particularly targeted design process.

A state-of-the-art review on previously developed frameworks supported the initial identification of the existing types and categories of measures (Matos Silva and Costa, 2016), a systematization process that, as envisioned, provided results of a general nature based on comprehensive research. As the next chapter discloses, adjacent to this systematization process, contextualized examples worldwide were gathered with the purpose of supporting the ongoing classifications with illustrations of concrete situations

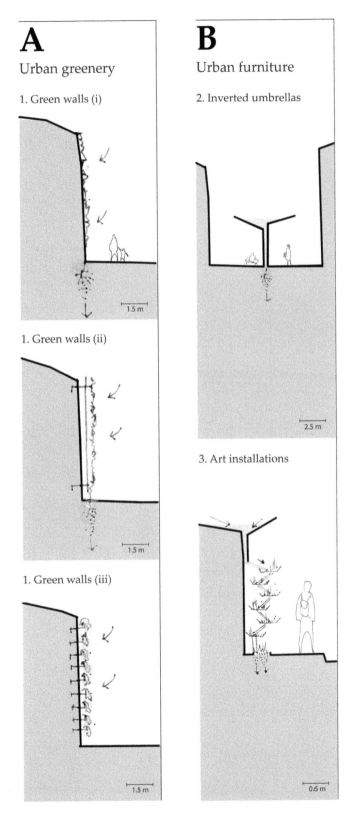

A

Urban greenery

1. Green walls (i)

1.5 m

1. Green walls (ii)

1.5 m

1. Green walls (iii)

1.5 m

B

Urban furniture

2. Inverted umbrellas

2.5 m

3. Art installations

0.6 m

Figure 2.14 Flood adaptation categories and types of measures applicable in the design of public spaces.

Source: Author's diagrams.

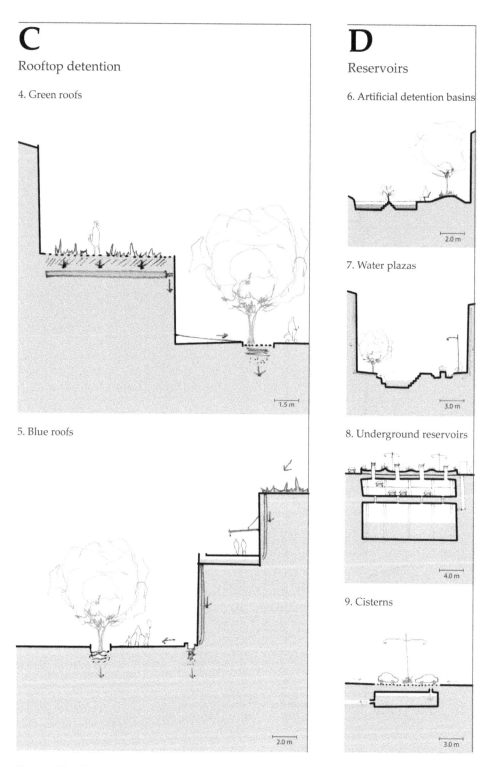

C

Rooftop detention

4. Green roofs

5. Blue roofs

2.0 m

1.5 m

D

Reservoirs

6. Artificial detention basins

2.0 m

7. Water plazas

3.0 m

8. Underground reservoirs

4.0 m

9. Cisterns

3.0 m

Figure 2.14   (Continued)

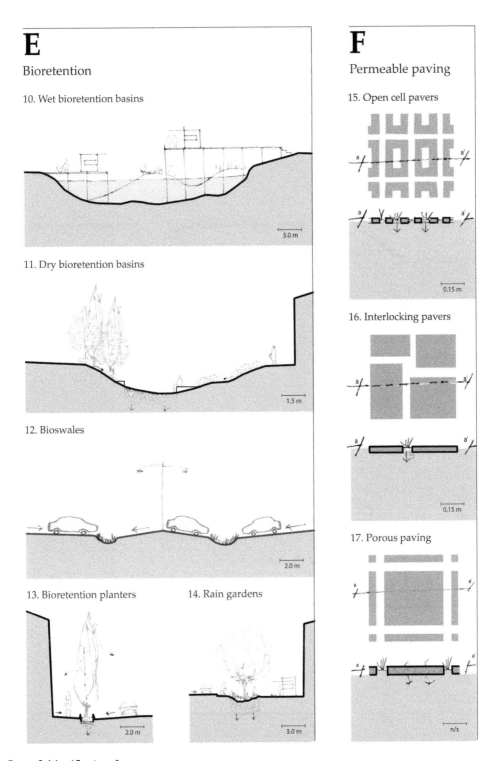

E

Bioretention

10. Wet bioretention basins

3.0 m

11. Dry bioretention basins

1.5 m

12. Bioswales

2.0 m

13. Bioretention planters

2.0 m

14. Rain gardens

3.0 m

F

Permeable paving

15. Open cell pavers

0.15 m

16. Interlocking pavers

0.15 m

17. Porous paving

n/s

Figure 2.14   *(Continued)*

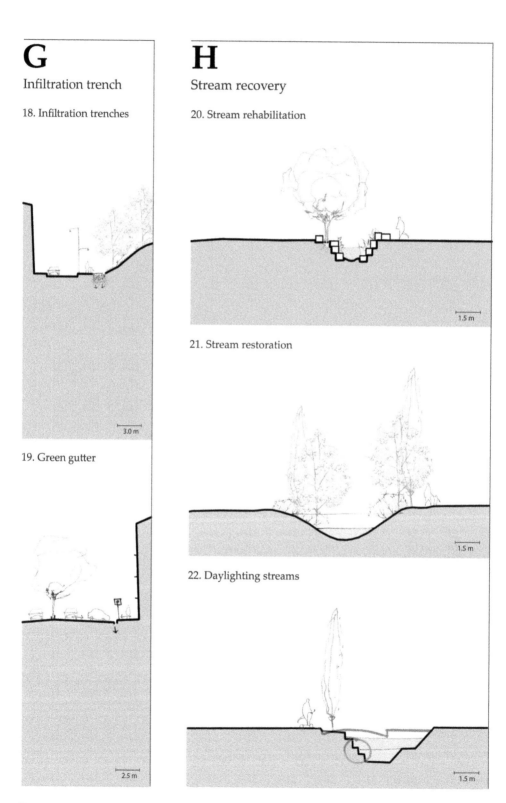

G
Infiltration trench

18. Infiltration trenches

3.0 m

19. Green gutter

2.5 m

H
Stream recovery

20. Stream rehabilitation

1.5 m

21. Stream restoration

1.5 m

22. Daylighting streams

1.5 m

Figure 2.14   (Continued)

I

Open drainage systems

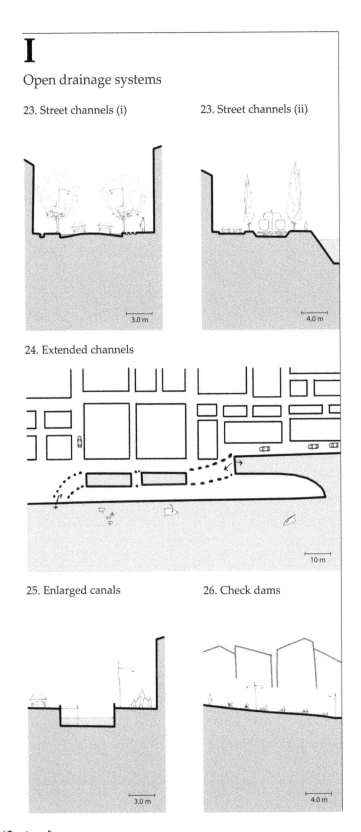

23. Street channels (i)

23. Street channels (ii)

3.0 m

4.0 m

24. Extended channels

10 m

25. Enlarged canals

26. Check dams

3.0 m

4.0 m

Figure 2.14 (Continued)

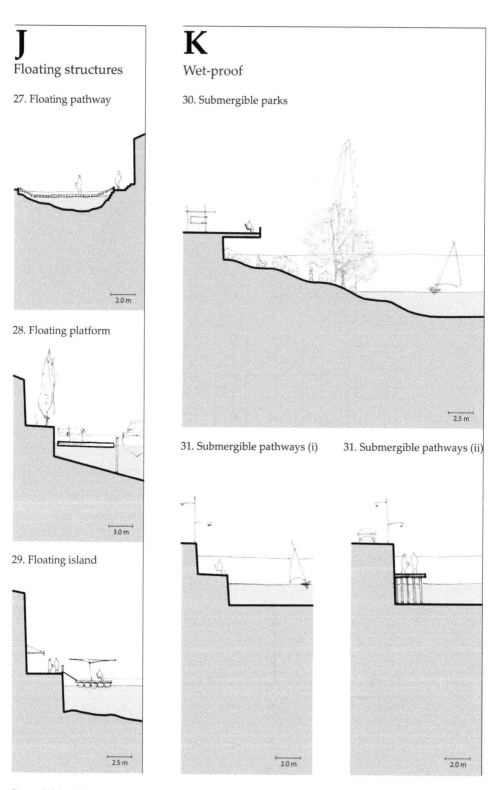

J
Floating structures

27. Floating pathway

2.0 m

28. Floating platform

3.0 m

29. Floating island

2.5 m

K
Wet-proof

30. Submergible parks

2.5 m

31. Submergible pathways (i)

2.0 m

31. Submergible pathways (ii)

2.0 m

Figure 2.14 (*Continued*)

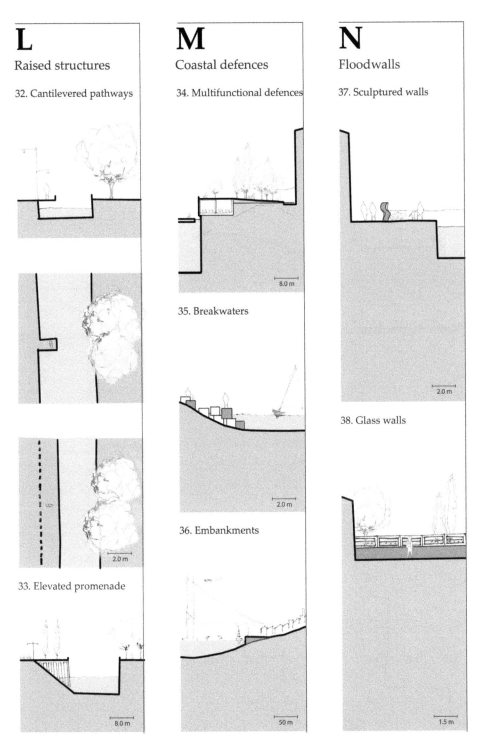

L

Raised structures

32. Cantilevered pathways

33. Elevated promenade

M

Coastal defences

34. Multifunctional defences

35. Breakwaters

36. Embankments

N

Floodwalls

37. Sculptured walls

38. Glass walls

Figure 2.14   (Continued)

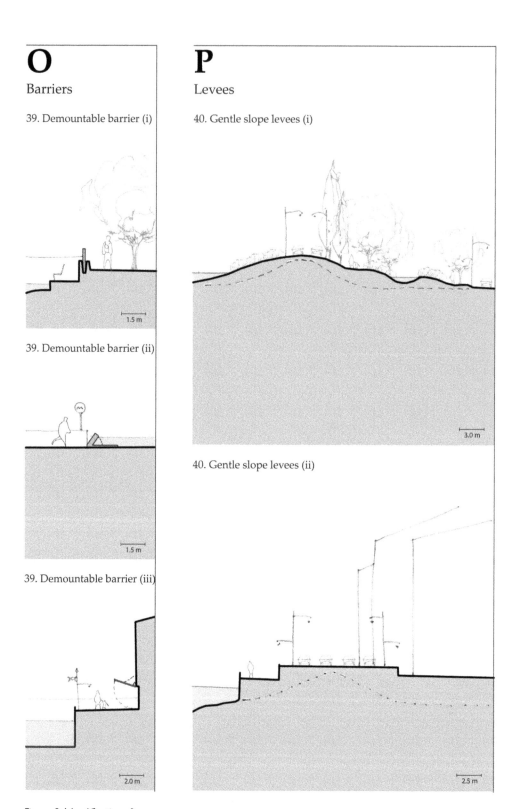

O

Barriers

39. Demountable barrier (i)

1.5 m

39. Demountable barrier (ii)

1.5 m

39. Demountable barrier (iii)

2.0 m

P

Levees

40. Gentle slope levees (i)

3.0 m

40. Gentle slope levees (ii)

2.5 m

Figure 2.14 (Continued)

specifically applied to public spaces. Ultimately, these two tasks endorsed each other, as the gathered range of examples also provided the identification of new types and categories of measures.

Efforts on climate change adaptation are multiple and have continuing developments and findings. As such, it is furthermore acknowledged that adaptation to urban flooding will continue to promote the development of new and relevant studies and, consequently, new and relevant types of flood adaptation measures applied in the design of public spaces.

Note

1 This research considered six infrastructural functions: harvest, store, infiltrate, convey, tolerate, and avoid. For more information regarding each infrastructural function, please consult Matos Silva and Costa (2016).

Chapter 3

Portfolio screening

Introduction

In the scope of the doctoral thesis that supports this publication, a database of examples was elaborated in order to follow the identification and characterization process of the types and categories of flood adaptation measures applicable in the design of a public space (full database may be consulted in Matos Silva (2016)). This specific analysis is based on comprehensive case studies highlighted in research projects and also on further bibliographical reviews, interviews with specialists, networking or in-site visits (Matos Silva, 2016, pp. 25–36). Besides including main or secondary functions related to flood vulnerability reduction, the chosen range of examples also aims to unveil intellectual treasures from existing designs and select "good quality" cases (PPS, 2003) among a comprehensive group of public space typologies (such as in Brandão, 2011b). The "portfolio screening" here presented is based on the developed empirical data collection focused on public spaces with flood adaptation purposes.

This portfolio screening is therefore not developed as an end in itself and should not be considered comprehensive but rather a significant sample of real-case situations that support and enrich the conceptual classification process. Albeit superficially, it further shows real-case examples that can be applicable to different contexts and with different purposes, hence providing further valuable information that may assist the decision making throughout the design processes.

I Caixa Forum square

- Location: Madrid, Spain
- Coordinates: 40°24'39.43"N – 3°41'35.36"W
- Construction date: 2006
- Design: Patrick Blanc
- Category, type of measure: Urban greenery; green wall
- More information: www.verticalgardenpatrickblanc.com

One of the façades of Madrid Caixa Forum entrance is a green wall implemented in 2006. This green wall is not built directly over the building's wall but rather in a supporting structure in order to avoid moisture problems. This structure, albeit attached to the building, is distanced enough so that it is possible to walk through its interior for monitoring of the

Figure 3.1 Green façade at the Caixa Forum square, Madrid, Spain.
Source: Author's personal archive, 2014.

irrigation and fertilization system. It is four stories high, and it includes over 15,000 plants from 250 different species, including *Arenaria montana, Bergenia cordifolia, Campanula takesimana, Cedrus deodara, Cistus purpureus, Cornus sanguinea, Dianthus deltoids, Garrya elliptica, Lonicera nitida, Lonicera pileata, Pilosella aurantiaca, Sedum alpestre, Yucca filamentosa,* and different begonias (Greenroofs.com, 2018).

This "living wall" not only contributes to the climatic amenity of the adjacent square but also to the reduction of the heat island effect, which is particularly present in a city such as Madrid that is warmed by continuous sun in the summer. Its capacity to harvest rainwater also contributes to the amelioration of floods, while the diversity of plant species constitutes an oasis for several types of birds and other animals.

Besides the environmental benefits, the design of the green wall may have additional aesthetic characteristics, particularly if considering it as an art project. In this example, designed by Patrick Blanc, one can identify a studied pattern of colors and textures that combine art, architecture and botany. For all these reasons, this square, highlighted by this wall, has currently become another drawing card in the heart of the Madrid's cultural district already surrounded by famous museums.

Other examples of public spaces next to the green walls of generous expression include the Musee du quai Branly in Paris or the Rubens Hotel in Victoria, London.

2   Expo Boulevard

- Location: Shanghai, China
- Coordinates: 31°11'9.45"N – 121°29'16.85"E
- Construction date: 2010
- Design: SBA Architektur
- Category, type of measure: Urban furniture, inverted umbrella
- More information: Knippers Helbig Advanced Engineering (2010)

The 2010 Expo in Shanghai, China, was held on both banks of the Huangpu River from 1 May to 31 October. Besides the China-Pavilion, the Expo Boulevard is the largest and most significant building that was kept on the site. One kilometer long and 100 m wide, it was the entrance to the Expo and the center of the major paths to its venues.

The Expo Boulevard extends across the former site to the banks of the Huang Pu River. Visitors are protected by a very large roof membrane. This unique roof is held by masts and six steel-glass funnels with a height of 35 meters and a free projection of 70 meters (Knippers Helbig Advanced Engineering, 2010).

The structure of the boulevard comprises four levels. The six funnels, or inverted umbrellas, are also called Sun Valleys as they direct the daylight into the basement floors. This network structure is further provided with an LED covering, producing

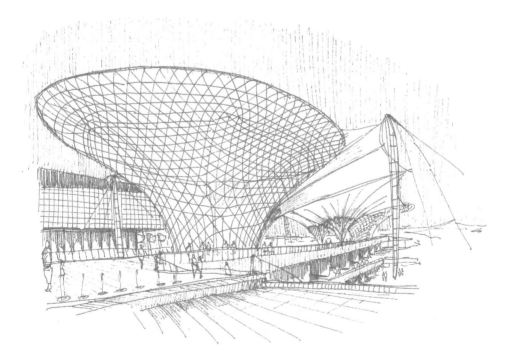

Figure 3.2 Expo Boulevard, Sun Valley – a series of six massive steel and glass funnels that collect water and sunlight, Shanghai Expo, 2010.

Source: Sketch by Vania Farinha, 2019.

dramatic light-changing effects at night. Yet the most important feature of this project in the scope of this analysis is the fact that the roof structure is also responsible for collecting rainwater, draining it into and toward the inner masts. This building with the form of an inverted umbrella is therefore here considered as a mega-water-collecting structure, formally defining the public space with distinct characteristics.

Other examples of water-collecting inverted umbrellas, on a smaller scale, include the ones existing in Woolworths shopping playground in Walkerville, Australia; the ones in North Road at Preston, UK; or the ones at the Taasinge Square at Copenhagen, Denmark (mentioned later in this chapter).

3 Jawaharlal Planetarium Park

- Location: Karnataka, India
- Coordinates: 12°59'3.93"N – 77°35'23.74"E
- Construction date: 2013
- Design: Vinod Heera Lal Eshwer
- Category, type of measure: Urban furniture, art installation
- More information: Mital (2013)

Together with the McCann Company, which funded this initiative, Vinod Heera Lal Eshwer built a permanent art installation at Jawaharlal Nehru Planetarium Park in India in order to raise awareness about the process of collecting rainwater by making it visible (Mital, 2013).

Located at the center of the children's park, this simple object clarifies the process of rainwater harvesting that is sometimes understood as a complicated term for the younger generations.

When it rains, a large funnel collects water into a see-through rectangular-shaped tank. Inside the tank there are some fish swimming, further recalling how life may be sustained by harvesting rainwater.

Visibly highlighted, the headline "catchtherain.org" is written in the tank. With this information, one may connect to a website with further information, namely a video and a free game for children, which reinforce the importance of collecting rainwater.

This example of urban furniture, which besides fulfilling its main purpose of communication and utility also contributes to flood adaptation purposes as it collects and stores rainwater, fits within the types of measures here considered as art installation and inverted umbrella. Other examples that more strongly relate to the type of flood adaptation measure of art installation include the works of Buster Simpson, such as the "Water Table/Water Glass" in Washington state and the "Whole Flow" in California.

4 Dakpark

- Location: Rotterdam, Netherlands
- Coordinates: 51°54'31.25"N – 4°26'17.00"E
- Construction date: 2009–2014
- Design: Buro Sant en Co
- Category, type of measure: Rooftop detention, green roof
- More information: www.santenco.nl

Figure 3.3 Permanent rainwater catching installation at Jawaharlal Planetarium Park in India.
Source: Sketch by Vania Farinha, 2019.

Dakpark is a roof park not far from the center of Rotterdam. The project program consisted of building a dike in the so-called Four Harbours strip that would also integrate offices, shops, schools and a public park. While the services were designed for the lower ground, the public park was planned as the roof of the whole structure. This project is thus representative of what it is meant by intensive spatial use or multiple ground use. Also fulfilling the function of a multifunctional dike, this roof park has a higher side that is around 8 m above the ground

Figure 3.4 Perspective of the Dakpark.
Source: Sketch by Vania Farinha, 2019.

level. The design further divides the park into two other intermediate levels. At its center, there is a generous stairway that connects the high and the low levels, with a water cascade.

Among the project's limitations are the constructional requirements associated with a green area being implemented over a roof and the attention that must be given to tree planting location or roof waterproofing. Another limitation is the safety concern related with its height. In accordance, the park encompasses a physical boundary of fences and gates and is closed off between sunset and sunrise.

5 Oliveiras rooftop garden

- Location: Praça de Lisboa, Porto, Portugal
- Coordinates: 41° 8'46.11"N – 8°36'54.12"W
- Construction date: 2013
- Design: Balonas e Menano, S.A.
- Category, type of measure: Rooftop detention, green roof

What was previously a standard square in downtown of Oporto, Praça de Lisboa, is now a commercial facility covered by the blanket of a garden. The facility, which is about one

Figure 3.5 Oliveiras garden green roof, seen from Torre dos Clérigos (the iconic tower and symbol of Oporto City).

Source: Author's personal archive, 2019.

story high, covers most of the square. Belowground it includes two levels of parking that existed prior to the square's rehabilitation. In its ground level it forms a semi-covered commercial street that connects two of Oporto's most representative icons, the Clérigos tower and Lello bookstore. At the level of the cover, the Oliveiras (olive tree) garden was designed, and an autochthonous meadow was planted alongside over 50 olive trees that aimed to relate to the city's ancient gates, formally known as Porta do Olival (Olive Grove Gate).

This project is an example of how a private facility may relate and dialog with its surroundings through an open and accessible public space that furthermore encompasses flood adaptation capacities. In similarity with other green roofs, the Oliveiras rooftop garden can retain an amount of rainfall volume that is significant, thus contributing to the alleviation of the drainage network.

There are countless examples of green roofs that are also public spaces. Among the many examples of green roofs, one must mention the Promenade Plantée in Paris, France; the High Line Park in New York, US.; Woman's University campus in Seoul, North Korea; the European Patent Office in Rijswijk, the Netherlands; or a more recent project such as the MAAT (Museum of Art, Architecture and Technology) building in Lisbon, Portugal.

6 Park Güell

- Location: Barcelona, Spain
- Coordinates: 41°24'49.91"N – 2°9'8.96"E
- Construction date: 1900–1914
- Design: Antoni Gaudí (Josep Mª Jujol y Juan Rubió i Bellver)
- Category, type of measure: Rooftop detention, blue roof
- More information: Francisco Caldeira Cabral and Telles (1999)

Park Güell is said to be one of the most notable and important works of Antoni Gaudí's career, as it is the most complete one (Ajuntament de Barcelona, 2018). It was built from 1900 to 1914; in 1969 it was declared a Cultural Property of National Interest, and in 1984 it was listed as a World Heritage Site by UNESCO.

The idea for the park, shared between the entrepreneur and industrialist Eusebi Güell and Gaudí, consisted in the creation of a residential area based on the garden-city movement, in similarity to those built in the second half of the 19th century in England. That is why the green areas were so important to include, and that is why the park is called "Park" Güell, in English. The urban development plan for 60 houses in 20 ha was not successful, partially due to the distance from the city center, which was significant in those days, and the fact that it was poorly connected with the adjacent territory. The urban

Figure 3.6 Detail of the hollow columns at Park Güell, through which water is conveyed from the terrace above into the belowground cistern.

Source: (imagIN.gr photography 2009)

development work therefore ceased in 1914, and in 1922 the municipality of Barcelona bought the property, turning it into a public park (Mailliet and Bourgery, 1993).

As we can all recognize, Gaudí's genius designing is strongly present in Park Güell. Faithful to his principles, he used the materials available on-site for the necessary construction, being very meticulous, at the same time, with the implementation of the street system and viaducts, studying the landscape in order to avoid unnecessary or excessive terrain modeling.

One impressive attribute of the park is its rainwater harvesting system. The residential area was planned to be included within a dense green park, yet the implementation area comprised the southeast slope of a hill known as "Muntanya Pelada" (Bare Mountain), which, as the name suggests, was very much depleted from vegetation in the end of the 19th century (Stefulesco, 1993). Considering Barcelona's precipitation indexes, the difficulty of bringing water from the urban network into the hill's high points and the poor water sources that existed within the terrain, one of Gaudí's challenges was therefore the irrigation of all the plantation works that were necessary. The solution was in the rainwater and consisted of the usage of the whole park area as a rainwater catchment area, integrating urban works with water management.

The main and biggest cistern of the park, with the capacity for 1,200,000 liters, is underneath the porch of the Nature Square (or Greek Theatre) over the Hypostyle Room (designed to hold a weekly market). The supply system of the cistern has been exhaustively studied. Water from the entire slope would reach the Greek Theatre square and there, through its porous pavement, infiltrate into the hollow insides of the neo-Doric columns of the Hypostyle Room and, from there, conveyed into the cistern. This way, Gaudí was able to ensure the maintenance of vigorous vegetation.

Other examples of blue roofs, as here considered, include the Stephen Epler Hall in Portland, US, or the Walterbos Complex in Apeldoorn, the Netherlands.

7 Parc Diagonal Mar

- Location: Barcelona, Spain
- Coordinates: 41°24'28.11"N – 2°12'49.34"E
- Construction date: 2002 (functioning since 2003)
- Design: Enric Miralles, Benedetta Tagliabue/EMBT arquitectes associats
- Category, type of measure: Reservoir, artificial detention basin
- More information: CLABSA, www.mirallestagliabue.com

Diagonal Mar Park, designed by Enric Miralles, is situated in the downstream end of Diagonal Avenue in Barcelona. More specifically, it develops within the interior of a residential block that is surrounded by the streets of Passeig de Taulat, Carrer de la selva de Mar, Carrer de Lull, and Carrer de Josep Plà. At more than 14 ha, it is the second-largest public park in Barcelona (Ajuntament de Barcelona, 2018).

It is a special park within Barcelona, as it is one of the few that uses groundwater for irrigation purposes (Vidiella and Zamora, 2011). It is considered by CLABSA (Clavegueram de Barcelona, S.A.), the company that manages Barcelona's sewerage infrastructure, as a "regulatory park," namely because of its constituting lake, with total capacity of 17,500 m^3 (CLABSA). This lake is compartmentalized into different storage tanks that enable a better control of water fluxes.

Figure 3.7 Detail of Parc Diagonal Mar, Barcelona, Spain.
Source: Author's personal archive, 2010.

The notorious and unconventional form of the lake, with abstract edges, occupies a significant area of the park. Besides its aesthetic and recreational features, this lake also functions as an uncovered reservoir for the storage of phreatic groundwater and rainwater that falls within the surrounding area. The lake, its ponds and the fountains use groundwater untreated with chemicals (not potable) so that aquatic habitat for fish, turtles, frogs, and birds can develop. Overall, planted vegetation demands little water, and some of the lake's margins encompass filtering wetlands, planted with native aquatic and riverine flora that facilitates water oxygenation and promotes biodiversity.

The use of pervious surfaces is also evident in this park, further sustaining an example of an ecologically sound urban park that is able to harvest and collect stormwater. Water is present in almost every part of the park, which furthermore includes several water features such as vapor sprays springing out of tubular sculptures.

Other examples of artificial detention basins include the case of Parc del Poblenou in Barcelona, Spain, or the ancient example of the Temple of Preah Vihear in Choam Ksan, Cambodia.

8 Benthemplein square

- Location: Rotterdam, the Netherlands
- Coordinates: 51°55'40.13"N – 4°28'36.29"E
- Construction date: Finalized in 2013

- Design: De Urbanisten
- Category, type of measure: Reservoir, water plaza
- More information: www.urbanisten.nl

With the same objective to retain stormwater for posterior gradual release of runoff and thus attenuate peak flows, new ideas arise within urban design practices and public space design. That is particularly the case of the Water Plaza in Benthemplein, Rotterdam, developed by "De Urbanisten" office, an urban research and design office based in the Netherlands. Coupled with strong creativity, it is the public space itself that incorporates the retention basin function in an urban context of intensive use. The total surface area encompassed within this project is 9500 m², including street and parking. The actual square-reservoir has an area of 5500 m² and offers 1.800 m3 of temporal water storage (De Urbanisten, 2013).

The idea of the Water Plaza had its root in the report "Rotterdam Water City 2005," a competition proposal that won the first prize during the Rotterdam Biennale in 2005 (Biesbroek, 2014, p. 119). Overall, this project combines synergies of flood tackling with public space renewal. Under this concept, Dutch cities such as Rotterdam can be divided into zones with autonomous watersheds where each zone is able to retain rainwater in its

Figure 3.8 Detail of the Water Plaza at Benthemplein, the Netherlands.
Source: Author's personal archive, 2014.

own public spaces through diverse conceptual designs (Boer *et al.*, 2010). It is usually considered as an efficient solution for compact urban areas where water does not flow away easily. Additionally, by making water visible, a closer social approach regarding the water cycle and flood dynamics is instigated.

This multifunctional public space is designed in accordance with very different quantities of retained water. While during most of the year the square will remain dry, if there is heavy rainfall, water will be temporarily stored within its interstitial spaces. Indeed, within this water square, some areas are deeper than others, and thus the flooding can be processed in stages; the borderline and smaller basins fill up first before the larger and deeper sports field. In this example, one can further note the attention given to the integration of open drainage and inflow channels as constituent elements of the public space design.

Finalized in 2013, the Water Plaza is considered a success when measuring international recognition and media attention. It is widely considered as an exemplary case of concrete climate change adaptation in a highly urbanized area. Another example here also considered as a water square is Tanner Springs Park project in Portland, US, designed by Atelier Dreiseitl and GreenWorks PC.

9 Escola Industrial

- Location: Barcelona, Spain
- Coordinates: 41°23'13.77"N – 2° 8'53.20"E
- Construction date: 1999
- Design: CLABSA – Clavegueram de Barcelona, S.A.
- Category, type of measure: Reservoir, underground reservoir
- More information: CLABSA

This underground reservoir is situated at the intersection of Viladomat and Rosselló streets, underneath the football field of the Escola Industrial de Barcelona. According to CLABSA's records, the reservoir has a total capacity of 27,000 m³, which is equivalent to ten Olympic swimming pools. Over the reservoir, a football field was installed covering more or less the same surface area of 94 by 54 meters with artificial grass. Among the infrastructural requirements, which needed to be included within the aboveground public space, are a ventilation system, an entrance to the machinery and control room and the necessary expansion joints. In this case, the ventilation system comprised chimneys, expansion joints were camouflaged below the grass and the entrance to the control room was made in connection to Carrer Rosselló. In this example, the entrance further included placards with detailed information about the technical characteristics of the reservoir as well as tri-dimensional schemes of its structure and integration with public space.

Other reservoirs were implemented throughout Barcelona, such as the cases of the reservoirs at Joan Miró Park, Bori I Fontestà garden, and Doctors Dolsa square, among others, all of which entail the common characteristic of including a certain type of public space over the reservoir, and thus conferring to this infrastructure the benefit of encompassing additional purposes such as squares, parking lots or gardens in addition to sports fields.

Figure 3.9 Detail of the football field area above the underground reservoir at Escola Industrial, Barcelona. Note the presence of the gray chimneys, part of the reservoir's ventilation system.

Source: Author's personal archive, 2011.

10   Parque Oeste

- Location: Alta de Lisboa, Lisbon, Portugal
- Coordinates: 38°46'46.53"N – 9°9'13.40"W
- Construction date: 2005–2007
- Design: Isabel Aguirre de Urcola
- Category, type of measure: Bioretention, wet bioretention basin
- More information: sgal.altadelisboa.com

The urban development of Alta de Lisboa neighborhood is bound by Parque Oeste (West Park). This urban park comprises a wet retention basin that, given the known lack of capacity of the downstream drainage network, essentially serves to regulate the increased amounts of superficial rainwater flows from the newly constructed developments. Besides the potential to store significant amounts of rainwater, this system also controls and reduces the velocity of the upstream flow, minimizing the influx at critical points. If

Figure 3.10 Detail of Parque Oeste/Oeste Park, Alta de Lisboa, Portugal.
Source: Author's personal archive, 2014.

provided with appropriate vegetation, the marginal areas of the basin would additionally serve for water purification.

The retention basin of Parque Oeste comprises a lake with 17,500 m² of maximum water surface. The limits of the lake consist of small "beaches" of grass or sand, concrete walls or gabions. From what it is known, no water is reused from the retention lakes. Up until now it could be said that it is a park with a very low affluence of users. However, this fact is more likely related with the complexities involving other urban matters (urban design, social matters and uncompleted neighborhood construction, among others) than to the design of the park itself.

Another example of a public space design project that includes a wet bioretention basin is, for instance, the case of Qunli park in Harbin, China, designed by Turenscape.

I I   Parque da Cidade

- Location: Porto, Portugal
- Coordinates: 41°10'9.20"N – 8°40'40.44"W
- Construction date: 1993
- Design: Sidónio Costa Pardal
- Category, type of measure: Bioretention, dry bioretention basin
- More information: Pardal (2006)

Figure 3.11 One of many dry bioretention basins at Parque da Cidade, Porto, Portugal.
Source: Author's personal archive, 2016.

Oporto's city park (Parque da Cidade do Porto) is located in a valley where several streams meet, the streams of Aldoar, Boavista and Nevogilde. From the highest point of "Avenida do Parque," it advances down to the Atlantic waterfront and its public promenade, making it particularly special and unique. It is almost the perfect location for an urban park. It was designed by Sidónio Costa Pardal, with a very clear inspiration from the English school of landscape architecture, with its simplicity and naturalist motivation and winding paths, and where different narratives and perspectives prevail over monotony and repetition.

It is said to be the biggest urban park in the country, but mostly, it is the pride of the city. It is a very popular park, frequented by people of all ages, who use it in a variety of ways: to run or ride a bicycle, to play, to walk, to have a picnic, to sunbathe and to read a book, among many other activities that can be seen at any time of year. The park also experiences different uses such as temporary fairs, academic or music festivals.

The design of the park includes a carefully chosen selection of species that, located in different areas, resist salty maritime winds and different types of solar exposure. *Cupressocyparis leylandii*, *Metrosideros excelsior*, and *Pinus pinea* are among of the most frequently seen species. Noise from the busiest street adjacent to the park, Av. Boavista in the south limit, is particularly well camouflaged by a significant terrain modeling together with a dense board of tree and bush vegetation. Also, the stone that came from the preexisting buildings within the park that were demolished was reused for the design of sustaining walls (built in dry stone), benches, tanks, ponds, shelters and false ruins.

One important characteristic of this park is the fact that, specifically through the design of its terrain grading, all pluvial water that falls within the park is retained and reused. The three lakes that configure the park, each one with its own hydrographic basin, are nourished by the water table that runs permanently toward the river. These permanent water levels provide rich conditions for heathy habitats and allow enough water resources for the park's irrigation.

Strolling throughout the park, one may further notice several small dry bioretention basins. These basins are composed by small concave depressions of large gravel (a loose aggregation of rock fragments) in order to promote quick and effective infiltration. Occasionally, these depressions may form small temporary lakes, and that is why they are in places that are not so busy. Some of these basins are partially surrounded by stone benches or sustaining walls. When dry, they can accommodate different uses. Quinta da Granja park in Lisbon, Portugal, can be considered as another example of a dry bioretention basin.

12   UMass Design Building

- Location: Amherst, Massachusetts
- Coordinates: 42°23'17.66"N – 72°31'25.29"W
- Construction date: 2017
- Design: Leers Weinzapfel Architects; Stephen Stimson Associates Landscape Architects
- Category, type of measure: Bioretention, bioswales
- More information: www.stimsonstudio.com

In 2017, the University of Massachusetts (UMASS) Amherst brought together the Departments of Landscape Architecture, Architecture and Building Technology programs in one single building. This building was named the "Design Building" and was envisioned as a space for interdisciplinary collaboration and experiment of shared sustainability principles. The Design Building and its surroundings are the first demonstration of such goals, with noteworthy applied sustainability practice. Built with CLT (cross-laminated timber) and glue-laminated columns, in contrast with the conventional energy-intensive steel and concrete buildings, the Design Building is already the recipient of several awards that highlight the best among good contemporary architectural examples.

The building was designed as a whole, including a collection of landscaped areas, yet most importantly to the portfolio screening here proposed is the way the design managed stormwater. The landscape architecture design responsible for this task, developed by Stephen Stimson Associates, encompassed rain gardens and bioswales with timber dams in the immediate surroundings of the building. The main idea consisted of directing roof and site runoff into the designed system of green storm management so that rainwater could be retained and filtered before it continues its path to the Connecticut river. Specifically, roof runoff is collected in a sculptural scupper that can be also viewed as a "spring source" at the top of the site (Figure 3.12 – right). In this vegetated system that embraces the building, rainwater flows slowly from the higher areas to the lower zones,

Figure 3.12 Left: Detail of the bioswale surrounding the University of Massachusetts (UMASS) Design Building. Right: Detail of the "spring source" at the top of the UMASS Design Building.

Source: Author's personal archive, 2019.

crossing stone beds and several check dams in order to reduce flow velocity and thus promote the deposition of solid debris. Upon reaching the low area, water may flood larger basins covered with riparian vegetation, which contributes to the filtration and infiltration processes. Surface dischargers connected to the conventional drainage system ensure that the designed flood levels aren't trespassed. The landscape architecture project was specifically recognized by the Boston Society of Landscape Architects (Merit Award 2018) and the Society for College and University Planning (Merit Award 2018).

Overall, the Design Building furthermore stands out for its educational purposes as a living laboratory about the built environment, supporting the curriculum and mission of the disciplines it welcomes (Stimson Studio, 2017).

There are several other examples of bioswales throughout the US, namely Portland, Seattle, Chicago, and Philadelphia, as well as in other countries such as England, Canada, or Germany, as it is an inexpensive measure to implement with immediate and evident results.

13 Taasinge Square

- Location: Copenhagen, Denmark
- Coordinates: 55°42'36.27"N – 12°34'4.53"E
- Construction date: 2013–2014

- Design: GHB Landscape Architects
- Category, type of measure: Bioretention, raingarden
- More information: www.klimakvarter.dk, www.ghb-landskab.dk

The Taasinge Square in Copenhagen is a great example of urban public space regeneration that transformed a leftover space, apparently useless, into a vibrant public square that is not only an ecological niche but also a best practice example on stormwater management.

Before this project, car transit and parking were the protagonists of the place. In 2014, the same year that Copenhagen was awarded as the European Green Capital, the designers of GHB Landscape Architects, among others, turned this site into a raingarden square that is able to manage 7000 m^3 of rainwater (Klimakvarter, 2019) while creating a place for community engagement.

The overall concept that led the project encompassed the idea of a climate change-adapted urban space, a space that would divert and reduce rainwater flows from its surrounding 4300 m^2 area into infiltration systems composed by dense and self-grown vegetation in order to delay its course into the orthodox drainage system. It is a vision that, among other social and ecological benefits, reduces the impacts of torrential rainfall that are expected to become more frequent and intense in the future.

Figure 3.13 Detail of Taasinge Square in Copenhagen, Denmark.
Source: Sketch by Vania Farinha, 2019.

Another aspect that must be mentioned about this intervention is its educational and demonstrative character. It includes sculpted elements throughout for children to play with, such as inverted umbrellas that catch rainwater and direct it into giant "drops" that serve as collecting reservoirs. From the "drops" reservoirs, rainwater can be pumped onto the pavement, provided with small channels that convey it into the raingarden. Furthermore, the raingarden is not considered as an isolated stormwater management infrastructure; on the contrary, it is a structuring and driving element of the square's integrated design, in which people can come close to, observe, smell and touch, through stairs and passageways. It is indeed a multilevel success story, excellent for a case study.

There are other examples of rain gardens integrated in the design of a public space, such as Edinburgh Gardens in Melbourne, Australia, or the Australia Road project in London, United Kingdom.

14   Can Caralleu

- Location: Barcelona, Spain
- Coordinates: 41°24'3.81"N – 2°6'48.02"E
- Construction date: 2006

Figure 3.14 Detail of Can Caralleu street. Different pavements were used. Almost half of the total area of intervention includes permeable pavement.

Source: Sketch by Vania Farinha, 2019.

- Design: Bagursa, Barcelona Gestió Urbanística. Sector Urbanisme
- Category, type of measure: Permeable paving, open cell pavement
- More information: Vidiella and Zamora (2011)

This project essentially consists of the enlargement of Carrer Major de Can Caralleu, turning a two-way road into a street with two downhill lanes and one uphill lane. Among its characterizing aspects are its situation of topographical asymmetry (steep road and bottom plain basin) and the use of permeable pavement as a design tool. More specifically, two types of porous pavement were used: grass, which naturally promotes the micro harvest, retention and infiltration of rainwaters; and gravel reinforced with recycled plastic cells over sublayers with different sizes of aggregates. The lower areas of the pavement function both as a parking lot and as rainwater retention areas collected through the permeable pavement's porosity. Under the parking area, in its lowest point, a cistern was additionally included. As the spillage risk of oils is high in parking lot areas due to the fact that vehicles may remain still for long periods of time, the proposed design furthermore works as an intermittent purifying filter based on the rain patterns. Overall, the permeable area corresponds to around 45% of the total area (4062.30 m^2), while the impermeable area corresponds to around 55% (5021.70 m^2) (Vidiella and Zamora, 2011).

15   Praça do Comércio

- Location: Lisboa, Portugal
- Coordinates: 38°42'27.05"N – 9°8'11.04"W
- Construction date: 2010
- Design: Bruno Soares
- Category, type of measure: Permeable paving, porous paving
- More information: brunosoaresarquitectos.pt, tecnovia.pt

Starting from 2008, the public limited company Frente Tejo, together with the municipality of Lisbon, were given the task to rehabilitate most of the city's waterfront. Within a global strategy, a major requalification was envisioned for Terreiro do Paço. This urban requalification envisioned key drainage retrofits, the rearrangement of traffic routes and the reassessment of street profiles as well as a renewed design for the central square.

Specifically in regard to the pavement design, the project envisioned an "earthen" material crossed by limestone stripes. According to the records of the project author, while the oblique perspective of the limestone stripes emphasizes the large size of the square and show the eccentricity of the statue, the earthen aggregate pavement and texture would recall memories of the square having been a Terreiro for 450 years. The aggregate pavement is thus constituted by limestone gravel and stone dust compressed with a colorless synthetic binder. In addition, it was further required for the pavement to have a high porosity rate so that it could quickly drain the excess of water from intensive rain episodes or storm surges. According to the building company, the implemented pavement has a porous surface and several drainage layers underneath. Between the bottom layer and the surface layer, there is a height of approximately 15.50 cm, which includes the regulating layer (also permeable) of 6.0 cm. This simple measure of pavement retrofit may prove to be of significant importance, particularly when considering that floods affect this square every

Figure 3.15 Permeable pavement initially composed of limestone gravel and a colorless synthetic binder, Lisbon, Portugal.

Source: Author's personal archive, 2014.

year. Other examples of porous paving include Greenfield Elementary and Percy Street, both in Philadelphia, US.

16 Elmer Avenue

* Location: Los Angeles, United States
* Coordinates: 34°12'39.39"N – 118°22'36.70"W
* Construction date: 2010
* Design: Stivers & Associates, Inc.
* Category, type of measure: Infiltration techniques, infiltration trenches
* More information: Robinson and Hopton (2011)

As part of the Los Angeles Basin Water Augmentation Study initiated in 2000, Elmer Avenue Neighborhood was selected as demonstration project in order to test and monitor state-of-the-art sustainable urban drainage systems (SUDS). Among the implemented measures along the sidewalks and private residential gardens are bioswales, permeable paving surfaces (including permeable concrete and permeable pavement), rain barrels

Figure 3.16 Elmer Avenue infiltration trench covered with water after rainstorm.
Source: Sketch by Vania Farinha, 2019.

and high-efficiency drip irrigation. Yet the most distinguishing feature of this project is the underground infiltration gallery below Elmer Avenue, which is capable of capturing 750,000 gallons of runoff (Robinson and Hopton, 2011). This project regenerated an important avenue of a neighborhood that initially had no flood management infrastructure and was thus vulnerable to recurrent flooding. Now, through an intervention area of 1.6 ha, this projected started to manage the first flush rainwaters from the surrounding 16.2 ha area. Through this public space retrofit, the aesthetic qualities of the street improved, increasing resident satisfaction with their block's walkability from less than 2% of survey respondents in 2006 to 92% in 2011 (Belden and Morris, 2011). Moreover, participation and project ownership were promoted through workshops, meetings, volunteer events and maintenance manuals.

17 Ribeira das Jardas

- Location: Sintra, Portugal
- Coordinates: 38°46'7.10"N – 9°18'2.34"W
- Construction date: 2001–2008
- Design: RISCO S.A., NPK Landscape Architecture
- Category, type of measure: Stream recovery, Stream rehabilitation
- More information: Programa Polis Cacém (2000); Brandão *et al.* (2018)

Figure 3.17 Detail of the stream rehabilitation project carried out at Ribeira das Jardas, Sintra, Portugal.

Source: Author's personal archive, 2018.

As a direct consequence of the promoted accessibilities to Lisbon's city center, such as Sintra railway and IC19 highway, Cacém became one of Lisbon's satellite suburban cities. Located in the middle of Jardas stream, it is also an urban area that is highly susceptible to recurrent episodes of floods (Pinho *et al.*, 2008). In addition, suffering from the lack of planning characteristic of the 1960s, Cacém was densely built with affordable housing, leaving little space for public use.

From 2003 to 2008, the downtown of Cacém was subject to an environmental and urban requalification project led by RISCO S.A. under the Polis program (Pinho *et al.*, 2008). Its main objectives focused on the improvement of mobility, identity, habitability, productivity and sustainability of this area (Programa Polis Cacém, 2000, p. 28). The intervention included the requalification of the Jardas stream through the creation of a retention basin as part of an extended linear public park alongside the water line. It also encompassed the regeneration of the surrounding public spaces inside the more intricate fabric, the restructuring of the road networks and the redesign of building areas (Programa Polis Cacém, 2000, p. 37). Ultimately, the project sought to promote a new centrality based in the linear form of the stream and its adjacent areas, connecting both margins and corresponding urban areas in a coherent system of public spaces.

Given the implications for public safety, the flood management of this urban area was considered a priority and as a result became the main generator of the whole project. In opposition to the mainstreamed approaches of flood defense that use rigid and extensive channel regularization measures, this project opted to increase the catchment and water retention capacity of the stream. Giving more space for the watercourse to flow, choosing to manage the risk of flood instead of controlling it, made this endeavor a pioneer intervention in Portugal at that time.

With the urgency and political will to face flood occurrences, the watercourse profile was therefore expanded through expropriations and demolition processes. Combining the need to resolve engineering problems of flood management with social needs for leisure and recreation and the goal to promote a new local character, this new linear public park embraced planning, engineering and landscape architecture, requiring an interdisciplinary design process that significantly contributed to the quality of the overall result.

Some infrastructural and ecological benefits brought by this intervention are evident: peak flow and flood occurrences were reduced, infiltration areas increased, biodiversity became richer – repercussions that are felt not only at the local level but also at the regional scale of the hydrographic basin. Among others, sewers were intercepted to new conduits parallel to the stream, native riparian vegetation was planted and, only when necessary, the stream was regulated with structural rigid measures.

Besides enhancing the continuity of the ecological network and the system of public spaces, other benefits of this intervention's design include the reduction of flood vulnerability through the engagement and awareness of actors with the natural hydrological processes; the exposure, within a multipurpose public space, of the Jardas stream as a unique and common value; and the consequent infrastructural monitoring by its users. In addition, Jardas linear park gave rise to an original centrality that may furthermore contribute with potential benefits such as the revitalization of the adjacent train station interface or even the emancipation of a new identity.

Among the range of examples of stream rehabilitation endeavors that encompass structuring public spaces, one must highlight the projects developed at river Volme in Hagen, Germany, by Atelier Herbert Dreiseitl, and the Catharina Amalia Park at Apeldoorn, the Netherlands, by OKRA Landscape Architects.

18 Eixo Verde e Azul

- Location: Jamor river basin, Metropolitan Lisbon, Portugal
- Coordinates: 38°44'51.24"N – 9°15'34.23"W
- Construction date: 2016–
- Design coordination: Amadora, Oeiras, Sintra and Parques de Sintra – Monte da Lua
- Category, type of measure: Stream recovery, Stream restoration
- More information: Matos Silva and Sousa Rego (2019)

The Jamor river basin is inserted in the hydrographic region of the Tagus (RH5) and comprises a total area of 44 km^2. It is furthermore integrated into the Regional Ecological Structure defined in the Regional Spatial Plan of Lisbon Metropolitan Area.

The main objective of the "Eixo Verde e Azul" project, currently under construction, consists of the requalification of the Jamor river basin through the regeneration of the public space adjacent to its main water lines. Specifically, it proposes the creation of an ecological corridor associated to a path of cycling lanes along the Jamor river, from its source at Serra da Carregueira to its mouth in Cruz Quebrada/Dafundo.

Figure 3.18 Work in progress. Detail of "Eixo Verde e Azul" stream recovery, next to Pendão
    aqueduct near the train station at Queluz, Portugal.

Source: João Sousa Rego, 2019.

The general design of the project was based in three strategic principles that translate a
new vision for the territory: the principles of continuity, of induction and of experiences.
The principle of continuity is associated with the continuity of water systems, biodiversity
and smooth mobility, thus promoting the movement and cohesion of living systems within
the territory. Through the induction principle, new relationships between patrimonial and
natural equipment are established, promoting the creation of structuring networks that
value existing assets. Finally, the experience principle is associated with the use of
public spaces along the streams, promoting different uses and functions such as recreation,
communal vegetable gardens, flood-prone areas, or amphitheaters carved into the hillside.

Unlike the most common actions in the territory, where each project, despite complying
with the provisions of a municipal master plan, follows a strictly local strategy, the "Eixo
Verde e Azul" project arises from a subregional vision of three municipalities, Sintra,
Oeiras, and Amadora, together with the company Parques de Sintra – Monte de Lua.
Among other benefits, this integrated strategy allows the compatibility of various functions
existing in the territory and ensures that some rely on the safeguarding of others. For
example, abandoned spaces encompassed rehabilitation projects that permitted a significant
increase of green urban areas, the guarantee of the maintenance and cleaning of the river or

the constitution of a large green park that encompasses a retention basin that mitigates the cyclical floods within the catchment area. Adjacent to this great valley is the National Palace of Queluz, which, due to its location, needs to become more resilient to floods and can now become a territorial pole adding a set of cultural and tourist facilities. Another example is the need to cross the barrier of the IC19 road, a difficulty that will be transposed by the construction of a green bridge that allows the reconnection between the gardens of the National Palace of Queluz to the woods of Matinha as well as provide a pedestrian crossing between upper Queluz and lower Queluz. Additionally, in the municipality of Oeiras, a hydraulic crossing is proposed over the A5 highway as a pedestrian crossing, allowing the connection between Serra de Carnaxide and the Jamor sports complex, creating green corridors that cross the council and allow daily movements of populations alongside public spaces.

Translating a vision of the territory that is supported by an ecological system of regional scale, further consubstantiated in the reinforcement of multifunctional corridors for the circulation of water, nature and people, in day-to-day life or in tourism, the "Eixo Verde e Azul" project contributes to the physical, economic and social regeneration of the areas it covers. The municipalities of Sintra, Oeiras and Amadora thus share a new landscape value and a new centrality, combining the rehabilitation of the National Palace of Queluz with its surrounding, giving back to the palace the capacity of generating pole of development of the region.

Other examples of stream restauration projects include the interventions made for Emscher Landscape Park in Germany or the near-natural restoration carried out at the Alb river in Karlsruhe, Germany.

19   Cheonggyecheon river

- Location: Seoul, South Korea
- Coordinates: 37°34'10.73"N – 127° 0'14.32"E
- Construction date: 2003–2005
- Design: Seoul City Government
- Category, type of measure: Stream recovery, daylighting streams
- More information: Kwon (2007); Novotny *et al.* (2010)

For over three decades, the Cheonggyecheon river was confined underground, over which passed a multi-lane roadway and an elevated highway. By the year 2000, strong structural fragilities of the speedway viaduct were identified. The costs for its recovery were considerable, and as such Seoul City Government considered the alternatives. In a political venture, Mayor Lee Myung-bak proposed not to invest in the renovation of the traffic infrastructure but rather on the restoration of the river's flow. In two years, the river was exposed and turned into a 5.8 km of linear park, which now crosses the city center. Among the resulting benefits is the improved capacity to sustain a flow rate of 118 mm/hr and flood protection for up to a 200-year flood event (Kwon, 2007).

What was formerly a source of congestion, pollution and aridity is now a blooming and environmentally healthy public space. Today, Cheonggyecheon river is a very popular park among the city residents, with clean water where people can swim and more than a few natural habitats. Sites of historic and cultural significance were also renewed, further contributing to a rehabilitated social identity.

Figure 3.19 Cheonggyecheon river promenade, Seoul, South Korea.
Source: Miguel Araujo, 2018.

More examples of daylighted streams that allow public space use include among others, the reconfiguration of several sunken canals in the Netherlands, especially in streets, squares or playgrounds – for example, the central Westersingel, a Rotterdam canal that, since 2012, can temporary store the extra water from severe storms on a lower-lying sculpture terrace prepared with flood-resistant urban furniture.

20   Banyoles old town

- Location: Girona, Spain
- Coordinates: 42°7'5.10"N – 2°45'53.48"E
- Construction date: 1998–2008
- Design: Mias Arquitectes
- Category, type of measure: Open drainage systems, street channels
- More information: www.miasarquitectes.com

Banyoles is a medieval town located in Catalonia, in between the Pyrenees and the Costa Brava. It was based in a very special area of karstic origin, near a natural lake formed in the Quaternary period (Brandi, 2014). Because this lake is topographically more elevated than

Figure 3.20 Detail of Banyoles old town public space. Girona, Spain.
Source: Sketch by Vania Farinha, 2019.

the town, the early settlement suffered from frequent episodes of flooding. In order to face this problem and manage the lake overflows, a system of drainage canals called the "recs" was built by the Benedictine monks in the ninth century. These canals, excavated out of the lacustrine travertine plate, run from the southeast bank of the lake, through the town and out into irrigation canals before reaching the river Terri. This unique drainage and irrigation system is Banyoles' first infrastructure and often matches existing property borders (Brandi, 2014).

In its first stages, the system of canals served to irrigate private kitchen gardens and public laundries. Later, the canals served the textile and agricultural industries as an energy source. By this time, the "recs" and its dynamics were an integrated part of city life. Yet throughout time, and as the industry changed, these canals progressively turned into sewage systems and therefore were gradually covered. Since then, and before the intervention coordinated by Mias Arquitectes, finalized in 2008, Banyoles lost all its connection with this ancestral system that was part of the town's identity.

The general program consisted of the requalification and update of the utility lines located underground and pedestrianizing almost all the area of the old Banyoles quarter. In line with the program, the project further sought to reestablish the town's essence and previous urban landscape uniqueness, with its water channels running through the streets and squares, offering the subtleties of freshness and sounds to passersby as well as a trip to the town's historical origins. Overall, new pedestrian areas were defined through the repaving of streets and squares in the center of the town, while the forgotten irrigation canals were reclaimed and occasionally uncovered. The project extensively explored the qualities of the travertine rock for the pavement and canal's design, through a detailed conception of its physiognomy, morphology and stereotomy. According to the authors

> The old town will now become a sequence of paths in which the inhabitants would have the possibility to enjoy the historical centre and its 12th century architecture. From now on, the pedestrian will always be accompanied by the presence of water.
>
> (Mias Arquitectes, 2008)

Although with conceptions based in quite distinct premises and programs, other examples of water channels integrated in the design of public space include Roombeek street in Enschede, the Netherlands, or the streets at the Solar City in Linz, Austria.

21 Pier Head

- Location: Liverpool, United Kingdom
- Coordinates: 53°24'14.47"N – 2°59'47.09"W
- Construction date: 2009
- Design: AECOM Design + Planning
- Category, type of measure: Open drainage systems, extended channels
- More information: AECOM (2016)

Figure 3.21 Detail of the extended channel at Pier Head, Liverpool.

Source: Sketch by Vania Farinha, 2019.

Pier Head is very important to Liverpool's sense of identity. In light of its status as a World Heritage Site and under the opportunity of Liverpool's 2008 European Capital of Culture, Pier Head embarked on an ambitious program to renew and regenerate its public space, more specifically, the central docks and the area in front of the Three Graces historic water-side buildings. Together with Liverpool Vision, AECOM coordinated the regeneration master plan, which encompassed the new landmark Museum of Liverpool Life, as well as a mixed-use development of homes, shops and offices along with a remodeling of the Mersey ferry terminal (AECOM, 2016). Yet the main focus of this prominent project was the creation of a channel extension linking the Leeds and Liverpool Canal to the north with dockland water basins adjacent to King's Waterfront to the south. It is the first major canal extension in the UK in a generation, with 650 m in length (Landscape Institute, 2014). With the Three Graces as a background, this 2.5 ha public square facing the river Mersey combined sunken water basins with open-air amphitheaters for cultural events. This public space was further designed to achieve a long life through the selection of robust materials and careful detailing.

22   Queen Elizabeth Olympic Park

- Location: London, United Kingdom
- Coordinates: 51°32'48.85"N – 0° 0'59.27" W
- Construction date: 2012
- Design: LDA Design with Hargreaves Associates (North Park)
- Category, type of measure: Open drainage systems; Enlarged canals
- More information: www.queenelizabetholympicpark.co.uk, www.susdrain.org/case-studies

The Olympic and Paralympic Games held in London in 2012 prompted the regeneration of a former industrial and commercial development at Stratford, East London, known to be a

Figure 3.22 Right: Queen Elizabeth Olympic Park enlarged channel: detail of the designed wetlands. Left: Queen Elizabeth Olympic Park: detail of a bioswale.

Source: Author's personal archive, 2017.

repository of rubble from the demolished buildings during the Second World War (Sus-drain, 2019). The site was furthermore contaminated by postwar munitions, batteries and waste from match-making factories, among others (Queen Elizabeth Olympic Park, 2018). Even after remediation projects, the implementation of infiltration measures was not advised, limiting the design options for sustainable drainage systems.

This urban regeneration project included the creation of a 102 ha park that was later named Queen Elizabeth Olympic Park. The landscape designers responsible for the North part of the park, namely LDA Design in collaboration with Hargreaves Associates, were able to convert this problematical site in an ecological public space system that con-tributed to the development of East London. A place that was previously a toxic waste bin was momentarily the ground of Olympic Games events and festivities and later an enduring new Public Realm landscape.

The design of the park took special attention to flooding and climate change projections, being able to manage floods with a return period greater than once in every 100 years (Sus-drain, 2019). This effort entailed significant terrain modulation (topography alteration) such as the enlargement of the river Lea channel, giving it more space to flow. This terrain alter-ation and channel extension further allowed the formation of a wetland bowl, which turned out to be a place not only for biodiversity enrichment, water quality and flood management,[1] but also a place for public use, for wandering through the existing paths or for staging shows in floating platforms. Indeed, the interdisciplinary ideals and consequent design of this park enabled the creation of places with a multifunctional character. Other drainage measures present in the park include bioswales, filter strips, or small bioretention basins Throughout the park, bioswales manage runoff from high to lower levels and into small wet bioretention basins. These wet basins, provided with sluices, manage water volumes, in order to attenuate peak flows while maintaining a permanent water level.

23   Kronsberg hillside avenues

- Location: Hanover, Germany
- Coordinates: 52°20'30.73"N – 9°50'23.01"E
- Construction date: 1998–2000
- Design: Atelier Herbert Dreiseitl
- Category, type of measure: Open drainage systems; Check dams
- More information: Dreiseitl and Grau (2005); City of Hanover Water Department (2000)

The Kronsberg development arose as a result of the 2000 World Exhibition. This new urban development was advertised as having been designed under the ideal of a comprehensive inte-gration between man, nature, and technology, in line with the theme of EXPO 2000 (City of Hanover Water Department, 2000). Although such a challenge was not accomplished in all the facets of an urban development project (Dreiseitl and Grau, 2005), it is clear how the stormwater management design is considered as having successfully fulfilled the expectations.

The main goal of the stormwater management project was to lighten the negative surface-sealing effects that a common urban development brings through the impermeabi-lization of land for buildings and roads on what was previously farmland. With the purpose of maintaining an undisrupted natural water cycle, traditional drainage systems, which rapidly convey stormwater into underground sewers, were called into question. The main objective therefore consisted of taking advantage of rain and all its ecological and social

Figure 3.23 Detail of the check dam system at the Kronsberg hillside avenue.
Source: Sketch by Vania Farinha, 2019.

potential, designing a stormwater management system that would work as closely and as similarly to nature as possible, "It was the idea from the start to slow down the rain's reaching the nearest drain. (…) The result was a close-to-natural rainwater system" (City of Hanover Water Department, 2000, p. 12). In order to accomplish this objective, multiple techniques, combining aboveground and underground components, were implemented, such as retention areas, rainwater reservoirs, bioswales, and check dams, among others.

The design of the northern and southern hillside avenues, with a slope of around 5%, are particularly relevant as an example of applied check dams in an open drainage system that forms part of an integrated public space. Because water hardly infiltrates in these areas, given its poorly permeable calcareous subsoil, the designers chose to combine other "ecosystem-based" stormwater management techniques besides the ones primarily targeted at infiltration. Rainwater that falls in each avenue is led into the center of the bioswale, composed by dense vegetation, several check dams and a network of paths, conveying water downhill through the surface and into retention basins. Throughout the course of water, one may find wooden and stone bridges, streams, stone benches and dammed water areas that occasional allow the permanence of small water ponds together with small cascades. As argued by Kathrin Brandt, "These water elements create a healthier microclimate, as water stabilises air temperatures and reduces dust clouds considerably" (2006, p. 7). Using photovoltaic pumps, rainwater from the downhill retention areas is sent to the top of the avenue so that the designed bioswale can have running water throughout the year and not solely in the rainy season. This option further encompasses educational purposes with its particularly noteworthy design "to give the future residents a vision of the importance of water" (City of Hanover Water Department, 2000, p. 14).

Bearing in mind the scale and moment of the Kronsberg development project (130 ha), it is moderate to state its ground-breaking contributions on urban hydrology and drainage, setting the path for the application of a new stormwater paradigm – a new paradigm that manages rainwater to the benefit of a healthier urban design, improving ecological systems, moderating the air temperature and creating quality public spaces for inhabitants.

Other examples of applied check dams can be seen, for instance, in the Renaissance Park in Tennessee or 21st Street in Paso Robles, both in the United States.

24   Yongning River Park

- Location: Taizhou, China
- Coordinates: 28°39'36.51"N – 121°14'53.08"E
- Construction date: 2002–2004
- Design: Turenscape
- Category, type of measure: Floating structures, floating platform
- More information: www.turenscape.com

Figure 3.24 The floating square over the wetland of Yongning River enables the fruition of the site during the flood season.

Source: Sketch by Vania Farinha, 2019.

Before 2002, the riverbanks of the Yongning river were made of concrete. In order to further control flood and stormwaters, the idea to completely channelize the river was under progress. Yet the local authority was persuaded by the proposition of a less costly and equally effective solution. The alternative solution comprised an ecologically sane while culturally and historically rich urban tidal park designed by Turenscape.

With the purpose of meeting the objectives mentioned, this project aimed at the confluence of two main systems: the ecological system that can serve floods and wildlife and the social system of public spaces that can serve people and tourists. While the ecological system essentially comprises the wetland and its inherent natural characteristics, in the social system the concept of a floating structure emerged. While the ground layer is frequently flooded for the benefit of its natural habitats and vegetation, sustaining up to a 50-year flood event, the floating "human layer" is composed of a path network that extends from the urban fabric down toward the park and a matrix of squares and groves of native trees. The enjoyment of this site is now available all year round through the use of floating structures, which further allow visitors to fully acknowledge the surrounding natural processes of seasonal flooding without compromising their safety.

Other examples of floating platforms that integrate public spaces include swimming pools, kiosks, sports fields, or bridges, particularly the pedestrian bridge at the London Docks connecting Canary Wharf with West India.

25   Bèsos River Park

- Location: Barcelona, Spain
- Coordinates: 41°25'25.14"N – 2°13'28.84"E
- Construction date: 1996–1999
- Design: Consortium for the Defense of the river Besòs, Barcelona City Council, City of Santa Coloma de Gramanet, Council of Sant Adrià del Besòs, City of Moncada i Reixac
- Category, type of measure: Wet-proof, submergible parks
- More information: Margolis and Robinson (2007)

Besòs River Park is located along the last five kilometers of the Besòs River before it reaches the Mediterranean Sea. It is one of the two rivers that bounds Barcelona's metropolitan area, the Lobregat river in the southwest and Besòs river to the northeast. Under the influence of a torrential rain pattern, characteristic of Mediterranean areas, water flows in the Besòs river typically shift from being reduced in the dry season into raging torrents in the rainy season.

After the severe floods of 1962, which caused around 800 casualties and widespread property damage (Margolis and Robinson, 2007), the river's walls were reinforced up to a 4 m height with concrete walls. Although such an approach was considered as the most effective flood control response, it reinforced this area as an unattractive space that became gradually neglected and environmentally deteriorated.

In 1996, the European Union sponsored the rehabilitation of the river Besòs banks, subsidizing 80% of the project's value under the Cohesion Fund. By this time the river was considered one of the most polluted in Europe (Margolis and Robinson, 2007). The project for the Besòs River Park was developed by the Consortium for the defense of

Figure 3.25 Detail of Besòs River Park.
Source: Author's personal archive, 2009.

Besòs river basin, formed by a collaborative process that involved government and citizen groups as well as intermunicipal cooperation agreements. Overall, the project recognized that the future urbanistic and ecological success of the river and its margins depended on the active use and care of its neighbors. As such, besides the objectives of improving wastewater treatment through the implementation of wetlands and expanding the hydraulic capacity of the river, it further entailed the creation of wet-proof public spaces for leisure and recreation alongside the river's margins.

During severe rain events, stormwater may rise up the channel walls and flood the park areas. In order to allow users a safe access to the flood-prone margins within the river channel, electronic placards were placed in the park's entrances. In accordance with the variation and severity of the stream flows, which is monitored through data collected by river-wide sensors, satellite and weather radar information, and river footage of the river-banks, this system of placards serves to inform potential users if they can or cannot enter the park. If the situation is dangerous, sirens and loudspeakers are also used. It is a partic-ularly efficient communication system, as it is able to rapidly inform people about sudden and intense flood surges (Prominski *et al.*, 2012).

In order to manage runoff flows, the innovative solution of inflatable dams was also implemented throughout the course of the Besòs river. This flexible infrastructure can rapidly inflate in order to detain water and, also in a matter of minutes, deflate in order to release water. This type of technology is particularly appropriate for small to medium-sized streams and is suitable for a wide range of purposes, from groundwater recharge to tidal barriers and recreation.

Other examples of submergible parks include, among others, the Rhone River Banks in Lyon, France; Parque fluvial del Gallego in Zuera, Spain; or Park Van Luna in Heerhugo-waard, the Netherlands.

26   Passeio Atlântico

- Location: Porto, Portugal
- Coordinates: 41° 9'56.62"N – 8°41'19.25"W
- Construction date: 2001–2002
- Design: Manuel de Solà-Morales and others
- Category, type of measure: Wet-proof, submergible pathways
- More information: www.manueldesola-morales.com

The waterfront renovation from Avenida Montevideu to Matosinhos is one of the most emblematic urban requalification projects of the beginning of the new century in the city of Porto, promoted by "Porto 2001 – European Cultural Capital." The project, coordinated

Figure 3.26 Detail of the submergible pathway at the Passeio Atlântico public space.

Source: Author's personal archive, 2006.

by the Spanish-Catalan architect Manuel de Solà-Morales, envisioned the creation of a wide linear public space that would serve as an intermediate space between the excessively ordered avenues upstream and the naturally irregular Atlantic coast with occasional scarps and rocky valleys. Overall, it intended to provide a change in the existing urban form through the introduction of a new way to perceive the morphology and dynamics of the coast (EU Mies Award, 2005).

One of the highlights of this project consisted of the renaturalization of the Galinheiras valley, reconfiguring the topography of the gorge's mouth as it was before the embankment and providing a viaduct as an alternative for car traffic. This great operation allowed a generous connection between the city park (Parque da Cidade) and the beaches, through a new landscape that integrates several natural systems and public space uses. At the south end of the viaduct, the Praça Gonçalves Zarco was transformed into a large roundabout by the coast with an underground car park in its center. This intervention of significant scale was not immune to the force of nature and, a few years after its construction, the parking lot had already experienced flooding (Corvacho, 2003).

Most importantly for the portfolio screening here envisioned, which in this section aims to highlight the wet-proof category and the type of measure of submergible pathways, this project also demonstrated how a coastal defense can be integrated within the design of a multifunctional public space, more specifically with the requalification of the Montevideu waterfront. This seafront project, besides configuring a coastal protection from the Atlantic urges, encompasses a two-level walkway, which comprises an upper path that connects a sequence of gardens and a lower path that goes along the rocks and, if the weather permits, allows a close relation to the sea dynamics. This lower path can therefore be considered a public space tolerant to occasional flooding by storm surges, a public space that is also an infrastructure of flood defense that allows passersby to monitor its state while strolling through rocky edges and sand beaches.

27   Elster millstream

- Location: Leipzig, Germany
- Coordinates: 51°20'00"N – 12°22'15"E
- Construction date: 1996
- Initiative: Neue Ufer (New Shores)
- Category, type of measure: Raised structure, cantilevered pathways
- More information: Prominski et al. (2012)

In the 1960s, the Elster millstream was conveyed underground, stripping the city of Leipzig from its former urban landscape of squares, pathways and homes by the riverside. Although the revitalization projects for the Elster millstream, as well as for the Pleiße millstream, were initiated in 1991, the continuation of the projects in the years that followed were greatly instigated by the "Neue Ufer" (New Shores) initiative, funded in 1996. Since 1996, it was possible to uncover around 1200 m of stream (Neue Ufer, 2013). Financial sources that allowed the regeneration project to be developed included fund raising, donations, floodwater protection resources and the city of Leipzig as well as financial corporations through town planning contracts with neighbors and adjacent owners of the river (Seelemann, 2015). Throughout the millstream, some areas were completely redesigned,

Figure 3.27 Uncovered Elster millstream with suspended pathways along its course.
Source: Sketch by Vania Farinha, 2019.

while in other areas a complete intervention was constrained by the availability of space, particularly along the busy roads previously implemented over the vault of the culverted course (Prominski *et al.*, 2012). In these situations, contemporary materials and know-how were used in order to provide raised structures, such as cantilevered pathways and floating piers, amongst others. In these more limited interventions along the millstream, the regeneration project had little ecological recovery, yet it greatly improved the surrounding urban spaces.

Other examples of public spaces that have included elevated structures over streams within their design, either cantilevered or elevated, include Terreiro do Rato in Covilhã, Portugal, or the waterfront promenade at Bilbao, Spain.

28   Barra do Douro North Jetty

- Coordinates: 41° 8'49.16"N – 8°40'31.05"W
- Construction date: 2004–2007
- Design: Carlos Prata Arquitecto and Fernando Silveira Ramos
- Category, type of measure: Coastal defense, breakwaters
- More information: www.carlosprata.com

Figure 3.28 Viewpoint at Molhe Norte da Barra do Douro, Passeiro Alegre, Porto, Portugal.
Source: Author's personal archive, 2019.

The primary purpose of this project, located at the north side of Douro's river mouth, encompassed the need to ensure navigation safety conditions in the entrance of the river. Conventionally, the program to build a jetty of considerable dimensions would have been undertaken solely by engineers, with a very specialized perspective on marine dynamics, coastal protection and navigability. Yet this project entailed further ambitions, namely the goal to integrate the necessary infrastructure as a constituent part of the "urban realm," as in public space that is owned and controlled by the city. Overall one can refer to the example of Barra do Douro Jetty ("Molhe da Barra do Douro") as a robust pier which is also a public space that combines aboveground benches and an interior area below for inside facilities.

With a built area of 9000 m², this linear corridor extends more than 600 m into the ocean. Pedestrians may stroll through the infrastructure, both aboveground or through an interior gallery that was envisioned to function as a restaurant and a passageway to the lighthouse on stormy days. Different situations mark this public space along its length, enhancing the experience of new views that connect the city to its sea. It is a living and lived in space, highly dependable on the dynamics of waters and used by people in multiple and reinvented ways.

29   Sea Organ

- Location: Zadar, Croatia
- Coordinates: 44°7'2.50"N – 15°13'11.39"E
- Construction date: 2004–2005
- Design: Nikola Bašić
- Category, type of measure: Coastal barriers, embankments
- More information: Stamać (2005)

Known by its inhabitants as the "stone vessel," the city of Zadar is configured within a peninsula by the Adriatic coast. During the Second World War, the city and its surrounding walls were fiercely bombed. During the later reconstruction work, a rather uninteresting concrete wall was built in the northwestern limits of the peninsula, paying little attention to the connection between the city and its surroundings.

In 2004, with the purpose of bringing more tourists to the city, investments were made in the design of a new port for cruise ships located at the north seafront. In the scope of this urban regeneration project, the marginal strip embankment that surrounds the city was redesigned, not only to facilitate the connection of passengers from the port jetty to the city's entrance but also in view of a new public space to be appropriated by the citizens of Zadar.

One of the most emblematic interventions within this renewal of the city's coast was a 70 m long staircase toward the sea that is also a music organ. This particular project, called

Figure 3.29 Detail of the Sea Organ design at Zadar, Croatia.

Source: Author's personal archive, 2011.

the Sea Organ, was designed by Nikola Bašić with the help of professor Vladimir Andročec (sea hydraulics consultant), Goran Ježina (organ craftsman artist) and Professor Ivan Stamać from Zagreb (musical tuning expert) (Stamać, 2005), and envisioned the possibility to intermingle public space design with a musical artifact fed by the natural forces of waves. As mentioned by Ivan Stamać, "the outcome of played tones and chords is a function of random time and space distribution of the wave energy to particular organ pipes" (Stamać, 2005). The arbitrary melodies produced by the organ therefore mirror the sea's strength: they can be either smooth or intense varying in consonance to whether the sea is calm or rough.

The Sea Organ, implemented in a slightly curved shoreline, is divided in seven sections of staircases toward the sea. The section that is further north encompasses eight stairs, and each successive section is one stair shorter than the previous. As such, the section that is further south only comprises two stairs. While the harmonies produced by the organ draw the attention of passersby, the formal configuration of the staircase design deceives the density of the new landfill in the northwestern corner.

Overall, the outcome of such a design sustains a public space that is frequently used by all, tourists and local community. In 2006, the Zadar Sea Organ won the European Prize for Urban Public Space in Barcelona (European Prize for Urban Public Space, 2006).

Another example of a flood protection infrastructural embankment that is also a densely used public space includes the design for The Hague, at Scheveningen, by the Spanish-Catalan architect and urban designer Manuel De Solà-Morales (Oorschot, 2013).

30   Blackpool seafront

- Location: Blackpool, United Kingdom
- Coordinates: 53°48'9.83"N – 3°3'24.10"W
- Construction date: 2002–2008
- Design: AECOM/Jerde Partnership
- Category, type of measure: Floodwalls, sculptured walls
- More information: AECOM (2011)

Before this contemporary intervention, Blackpool seafront was composed by traditional defensive coastal infrastructure such as steep embankments and tall vertical walls. Connections to the sea or entrances to the beach were very few and made out of narrow and precipitous stairways.

The Blackpool Promenade project regenerated this seafront with the multiple purposes of responding to the threats of sea level rise and more frequent and extreme storm surges and of providing an extended and improved waterfront public space. Reinventing the traditional concept of floodwall, this 8 ha project integrates infrastructure and art while connecting people throughout and across the designed space. In a design that conceptually attempts to mimic nature through its undulating form, concrete stairs separate the town from the water and fulfill the apparently contradictory goals to protect people from and connect people to water. Differentiating itself from the traditional approach to coastal defense, this project therefore intentionally allows part of the public space to be flooded during extreme events.

Figure 3.30 Coastal defense in Blackpool, United Kingdom.
Source: Sketch by Vania Farinha, 2019.

31   Wells Quayside

- Location: North Norfolk, United Kingdom
- Coordinates: 52°57'28.41"N – 0°51'2.85"E
- Construction date: June 2012
- Builder and sponsor: Flood Control International and the Environment Agency (EA)
- Category, type of measure: Floodwalls, glass walls
- More information: www.floodcontrolinternational.com

In 2012, a new flood barrier, comprising a structural glass wall, was placed at the Wells Quayside in North Norfolk, UK, in order to avoid disastrous flooding events such as the ones previously experienced in many of East Anglia's coastal towns in 1953 and 1978 (Steers *et al.*, 1979). Before this date, oak flood boards were placed on top of the marginal low brick walls every October in order to increase the defense height and sustain the periodic river flooding. Every April, these timber walls were removed so that the views and connection to the river could again be reestablished.

The implementation of a permanent glass wall offers a solution that maintains the visual connection between both sides of the flood defense infrastructure. In addition, it offers the possibility for local awareness and community engagement in the sense that it allows a closer contact and consequent acknowledgment of the natural processes of river dynamics.

This 1m high glass barrier is said to protect more than 500 properties at the cost of £116,000 (BBC, 2012), developing along 80 m plus 18 m of sliding gate (Flood Control

Figure 3.31 Glass flood wall by the marginal road of Wells Quayside in North Norfolk, United Kingdom.

Source: Sketch by Vania Farinha, 2019.

International, 2019). Existing post fixings were reutilized for the support of the new panels, each approximately 4.2 m in length and 1m in height. According to the builder company, Flood Control International, these structural glass walls were designed in order to allow impact loading as well as full hydrostatic pressure, being furthermore advertised that they are quick and easy to install and require minimum maintenance when treated with a specific self-cleaning coating (Flood Control International, 2012).

Other examples of glass flood walls applied in public spaces can be seen in Dassow, Germany; Decin, Czech Republic; or in Northwich and Leeds, both in the UK (IBS Technics GmbH, 2019).

32 Kampen waterfront

- Location: Kampen, Netherlands
- Coordinates: 52°33'31.88"N – 5°55'1.41"E
- Construction date: 2001–2003
- Planning and construction: City of Kampen
- Category, type of measure: Barriers, demountable barriers
- More information: Voorendt (2015); Prominski *et al.* (2012)

Figure 3.32 Lowered down flood gate crossing a small street to be raised vertically when needed.
Source: Sketch by Vania Farinha, 2019.

Kampen city, located at the lower reaches of the river IJssel, is an old Hanseatic city in the Dutch province of Overijssel. In Kampen, the combination of strong winds and floodwaters flowing downstream can make the sea level rise up to 3 m in just three hours (Prominski *et al.*, 2012). In order to protect the town from flooding and, at the same time, maintain the city's traditional character, various measures have been implemented, integrating flood defense with other purposes. For example, in order to provide extra height, "stoplogs" were integrated in the existing quay wall. The historic city wall, as well as several private gardens and buildings, were also improved with additional demountable barriers (Voorendt, 2015). In an emergency setting, barriers are raised and gates are closed, specifically with mobile aluminum elements. These elements are stored in a large warehouse and can be totally installed by a 200-person-strong floodwater team within three hours (Prominski *et al.*, 2012). In some situations, the flood defense line crosses small streets or bigger roads. Overall, Kampen's flood defense strategy is completely integrated within its public spaces, which continue to provide other uses and purposes such as transport, parking and historical identity, among others.

33   Corktown Common

- Location: Toronto, Canada
- Coordinates: 43°39'14.60"N – 79°21'6.12"W
- Construction date: 2006–2014
- Design: Michael van Valkenburgh Associates, Arup
- Category, type of measure: Levees, gentle slope levee
- More information: mvvainc (2012)

Corktown Common is a 6.5 ha park in a postindustrial area left as a brownfield. Located in the West Don Lands district of Toronto, this park currently offers a wildlife-filled marsh, athletic fields, playgrounds and a generous area of open lawns (WATERFRONToronto, 2009). Yet this park is also a flood defense infrastructure. Designed by Michael van Valkenburgh Associates, partnered with Arup, the Corktown Common Park not only serves residents across Toronto but also encompasses a levee whose structure is hardly distinguishable. According to the designers, the levee embedded within the park is robust enough to protect vulnerable areas against a 500-year flood (mvvainc, 2012).

Placed over gentle slope banks, this levee-park protects the new West Don Lands community from flooding, including Toronto's financial district. Although it is a constructed landscape, it is particularly rich ecology-wise, with a great number of native trees and scrubs and an extensive marsh with developing and enriching biodiversity. This project further stands out for its flood management approach, which enables the collection and treatment of rainwater within the constructed marsh, subsequently storing it for irrigation (mvvainc, 2012).

Figure 3.33 Detail of the levee at Corktown Common Park in Toronto, Canada.
Source: Sketch by Vania Farinha, 2019.

Discussion

Climate change adaptation is still faced with numerous challenges. Most common barriers to adaptation can be associated with "short term thinking of politicians and long term impacts of climate change," "little finance reserved/available for implementation," "conflicting interests between involved actors," "more urgent policy issues need short term attention," or "unclear social costs and benefits of adaptation measures" (Biesbroek, 2014, p. 139). However, every day, successful examples grow in number, and as argued by Howe and Mitchell, it is increasingly important to see more empirical studies of adaptation examples rather than just dwell on the barriers to change (2012, p. 38). This emergent tendency of new and innovative adaptation projects can be exploited as a creative laboratory that proposes, assesses and monitors solutions through an ongoing learning process in order to serve and inform future decisions and reduce generalized hindering constraints.

For Jan Jacob Trip, public space may be the element of urban development that is most difficult to plan and design as it relates to so many intangible qualities inherent to the quality of the place itself (Trip, 2007, p. 81). Although there is no imposing formula that would define the quality of a public space design, it is commonly accepted that a good design must develop from a sensible understanding of its situation and all its encompassing contexts: environmental, cultural, social, economic and political (Brandão et al., 2002, p. 18). For Borja,

> the quality of public space can be largely evaluated by the intensity and quality of the social relations that it generates, by the force to encourage the mixture of groups and behaviours, and by the ability to stimulate the symbolic identification, expression and cultural integration.[2]
>
> (Borja, 2003, p. 124, author's translation)

A good public space design therefore is likely to result in a place that is valued and used and that stimulates a communities' sense of belonging (1996[1960]). Based on an exhaustive evaluation of thousands of public spaces worldwide, the nonprofit organization Project for Public Spaces (PPS) evidences that "great places" generally share four principal attributes, namely, sociability, uses and activities, access and linkages and comfort and image. PPS further developed "The Place Diagram" as a tool to assist the analysis of any place, good or bad. Intended as an all-inclusive generalist approach, the diagram also evidences the "intangibles" and "measurements" inherent to the presented principal attributes. Great public spaces that entail flood adaptation measures are likely to include at least one of the PPS intangible qualities of being "vital," "useful," "sustainable," or "safe." The full range of empirically collected examples (which can be consulted in Matos Silva, 2016) was based on the aforementioned references and emphasized attributes of quality public spaces.

The presented group of evidence further aimed to encompass a comprehensive range of public space typologies. For this purpose, the typology of public spaces identified by Brandão was used, namely the differentiation regarding "Layout spaces" (squares, streets, avenues), "Landscape spaces" (gardens, parks, belvederes, viewpoints), "Itinerating spaces" (stations, interfaces, train-lines, highways, parking lots, silos), "Memory spaces" (cemeteries, industrial, agricultural, services, monumental spaces), "Commercial spaces" (markets, shopping malls, arcades, temporary markers, kiosks, canopies), and "Generated spaces" (spaces generated by buildings: churchyards, passages, galleries, patios; by equipment's: cultural, sports, religious, children's; by systems: communication, lighting, furniture, art) (Brandão, 2011b, p. 35).

In several of the presented examples that enabled this analysis, adaptation measures were unrecognized as such. The existing functional qualities of some cases were rather associated to other, more prevailing, conceptual approaches such as sustainability or flood protection. Yet, bearing in mind the understandings around the concept of adaptation (Matos Silva and Nouri, 2014), all presented examples are considered as adaptation measures. Not only do all examples entail the transposition of uncertainty and its apparent impediments into public spaces of multifunctional qualities and uses, but also all examples serve as solid grounds for the assessment of adaptation action. Altogether, the examples provided are not meant to offer an exhaustive collection but rather a significant sample of designed solutions that endorse further reliable research and decision making in climate change adaptation endeavors.

Note

1   This included several other sustainable drainage measures such as bioswales, filter strips, or small bioretention basins. Indeed, throughout the park, bioswales manage runoff from high to lower levels and into small wet bioretention basins. These wet basins, provided with sluices, manage water volumes, therefore attenuating peak flows while maintaining a permanent water level.
2   Original text: "la calidad del espacio público se puede evaluar sobre todo por la intensidad y la calidad de las relaciones sociales que facilita, por la fuerza con que fomenta la mezcla de grupos y comportamientos y por la capacidad de estimular la identificación simbólica, la expresión y la integración culturales."

Chapter 4

Discussion

Floods are among the most frequent human-enhanced urban hazards, albeit paradoxically experienced as an exceptional event attributed to nature. Supported by extensive literature, climate change projections estimate an increase in the frequency and intensity of floods. Furthermore, public spaces are among the most vulnerable areas to flooding, as these are where impacts are more acutely experienced. Approaching the identified concerns as challenges rather than hindering constraints, the research supporting this book reinforces the premise that the design of public spaces is a key component of the urban adaptation to current and expected flooding events.

The climate change adaptation agenda has been creating revolutions that go beyond the restrictive arenas of climatic science, specifically within political arenas, flood risk management and public space design. In light of a deepened analysis concerning the concept of climate change adaptation, hereinafter referred simply as adaptation, it was possible to understand its meaning as a continuous learning adjustment process, which aims to reduce climate related vulnerabilities while seeking the advantage of beneficial opportunities. In contrast with sustainability planning, which supports decision making by analyzing past trends, adaptation processes base the options upon the recognition of possible future climates through projections and simulations. Adaptation therefore embraces the multiple possibilities of learning with mistakes. The city – and more specifically its public spaces – is extraordinarily adaptable, however, under a pattern of relatively stable changes. When facing unprecedented and potentially extreme changes, public spaces may not have the same autonomous adaptation capacity. It is in this context that planned adaptation gains strength against "business as usual."

Supported by literature, there is an increasing concern on the need to act in the face of uncertainty. Adaptation therefore presents itself as an instrument that helps the management of uncertainty – not only the uncertainty that is natural to the evolutionary processes of any city but also the uncertainty of future climate-driven projections. Both uncertainties were once thought of as unattainable, but their inclusion in planning practices is now widely discussed. In line with an increasing ambition to face the urban impacts stressed by climate change research, several countries, such as the Netherlands, United States, or the United Kingdom, have undergone regional and local adaptation undertakings. As is consensually agreed and recognized in literature, adaptation is prompting new urban planning approaches. Beyond this finding, it has been evidenced that public spaces can lead effective adaptation undertakings that are explicitly influencing urban design practices as we know them.

Specifically, regarding the recurrent phenomenon of urban flooding, climate change research has been warning about the fact that traditional flood risk management practices must be reassessed, particularly if projected impacts are to be managed, such as the likely increased frequency and greater intensity of storms (precipitation and storm surges) together with a rise in sea level. When analyzing an overview of the flood risk management practices throughout history, it was possible to further verify an emerging change, from the conventional focus from the goal to reduce the probability of experiencing floods to the aim to reduce society's vulnerabilities. As an increasingly discussed perspective, this notion will inevitably change the relationship between the city and (its) water. In addition, it was possible to highlight how the goal to tackle social vulnerability might be related to the management and integration of risk and uncertainty in flood management practices, notably by fully acknowledging and welcoming the processes of the natural water cycle amid public spaces. However, this is an argument that calls for improved knowledge. By making climate change visible, namely by incorporating the natural processes within the design of a public space, the social impact may have opposite repercussions and rather lead toward increased vulnerability, a fact that is particularly associated with how people and communities perceive and sense the risk of flooding.

Within a number of intrinsic roles, such as being a civic place of social and economic exchanges or a gathering place where cultures mix, public spaces have found an enhanced protagonism in light of the recognized need of a change of paradigm in current flood risk management practices, a fact that is demonstrated by cities that have matured their relationship with water, especially through the design of public spaces that, by recognizing water's bountiful and resilient capacities, promote a proximity between society and flood risk management infrastructure.

Through the inclusion of flood adaptation measures within public spaces, reinforced challenges arise before contemporary urbanism and urban design practices. Among those challenges are: (1) the acknowledgment that total flood protection is unrealistic and unwise; (2) a need for "out of the box" thinking; (3) a requirement to abide with further expanded horizons, whose actions must be periodically monitored and revisited; (4) the compliance with the notion of public space as an urban system; (5) the articulation with other municipal agendas or ongoing programs; and (6) a search for a combination of multiple strategies and solutions. In accordance, public space design has found new horizons of multidisciplinary and interdisciplinary practice, which may instigate further creativity, new horizons that will not only bring new approaches and technologies into the frontline of innovation but also stimulate the reencounter of culture and tradition.

One may furthermore argue that competent and politically autonomous municipalities that are close to their citizens are more likely to conduct effective adaptation action. Local initiatives such as specific public space interventions can raise the standard for good quality cities. The quality of future cities can therefore be influenced by the quality of adaptation measures applied in public spaces. Indeed, through public spaces, extended opportunities for experimental learning and monitoring, inherent to adaptation processes, are provided without neglecting values such as local identity or sense of place. Overall, it is argued that through public spaces, which provide the opportunity to integrate and reveal the complex connections between natural, social and technical processes, and specifically through public space design, traditional flood management practices may be enhanced can rise to the contemporaneity of our time.

In light of the identified examples of public spaces with flood adaptation purposes, it was possible to identify some of the potential benefits that may specifically arise from the inherent characteristics provided by public space. Among them are (1) the favoring of interdisciplinary design, (2) the embracement of multiple purposes, (3) the promotion of community engagement, (4) the support of an extensive physical system, (5) the opportunity to expose and share value and, finally, (6) the prospect of promoting risk diversification and communal monitoring.

After recognizing (1) the enhanced protagonism of public spaces in light of the paradigm shift in current flood risk management practices as well as (2) the potential benefits that may specifically arise from the intrinsic characteristics provided by public space, it becomes necessary to explicitly identify how public space can accommodate flood adaptation measures. Through a particularly targeted literature review, together with a semantics analysis, it was possible to identify and systematize 40 types of flood adaptation measures applicable in the design of public spaces, covered by 16 categories.

The result of thus systematization process is targeted at facilitating and accelerating the initial stages of a public space design project with flood adaptation ambitions, notably by exposing an extensible body of available options. Developed with the purpose to offer a commonly used vocabulary and simple technical notions, it further aims to support and promote communication and exchange of know-how. It furthermore acknowledges the advantageous possibility to add new knowledge as it becomes available. The proposed systematization process is therefore an unlimited work in progress, prepared to evolve and to be restructured in light of new teachings, concepts or approaches. Although the categories and types of measures here highlighted can provide a very useful starting point, most likely, and most fortunately, they will also change and develop as new challenges arrive. Prompted by the urging need to adapt our cities when facing potential climate change and unprecedented flood events, the proposed flood adaptation measures applicable in the design of public spaces offer a different approach to tackle the well-known problem of urban flooding. Through a different perspective, one that highlights the importance of public space design in adaptation undertakings, this book presents a specific group of measures that confront traditional flood risk management practices. Through the design of public spaces with flood adaptation capabilities, our urban territories can become better adapted for the present and projected flood impacts.

While the most attractive adaptation strategies are usually those that offer development benefits in the short term and reductions of vulnerabilities in the long term, extensive literature has been highlighting that not all adaptation responses are benign. Selecting the optimal adaptation strategy or measure for a particular situation is neither easy nor straightforward. This determining process is particularly complicated given the specific ramifications and secondary impacts related to adaptation processes. In some cases, results can be critical, namely when adaptation does not fulfill its designated objective and ultimately leads to increased vulnerability. This phenomenon is generally called maladaptation. However, maladaptation cannot be considered as a hindering factor supporting "business as usual," as uncertainties can be minimized through the ongoing adjustments of continuous assessment. Facing an unprecedented area of action, concepts, paradigms or structures are expected to change over time, as are the functions, appearance and complexities of public spaces with flood adaptation measures.

Another prominent objection to the advocated inclusion of flood adaptation measures within public spaces is the assumed reliance on local-scale social response. Regardless, although limited confidence is provided on what extent local empowerment can favor or harm "bottom-up" adaptation processes, it has been evidenced that positive outcomes arise when the design of a public space with flood adaptation capacities involves local people and communities. Moreover, people and communities are not only targets but can also be active agents in the management of vulnerability. This potential propensity to facilitate adaptation action can be instigated through the design of public spaces that make climate change visible and hence meaningful. Yet by providing an additional source of knowledge through the design of a public space, counterproductive misunderstandings may also occur. By questioning the certainty of local-scale social response, the advocated need for continuous monitoring and learning alongside adaptation undertakings is reinforced.

It is furthermore essential to highlight that, while the analyzed initiatives have counterbalanced the inevitable uncertainties of global models and the generalized "top-down" policies, local action must be connected to global scientific findings and their encompassing strategies. In general terms, one can note that if science and "top-down" approaches can provide important knowledge, then local "bottom-up" approaches can provide critical wisdom. Nevertheless, both approaches are important and must not be isolated. If local adaptation is not associated with a bigger strategy, it will get lost in scale and lose its strength, and if "top-down" approaches are distanced from specific local hindrances, they can fail to reach communities and fail to tackle the most prominent vulnerabilities. While local-scale action is presently acknowledged as a fundamental element for effective urban climate change adaptation, its greater challenge relies on finding the balance and exploring the benefits from the arising synergies between local collective actions and national and international strategies. The same way local adaptation strategies must not be dissociated from global adaptation strategies, so too with the processes of public space design, which must follow objectives and strategies of regional and national levels.

Flood risk management in particular has been essentially controlled by specific technical and specialized disciplines that have authoritatively decided upon the actions required with regard to coastal, riverine or pluvial flooding. Yet a different perspective is here proposed, one that highlights the importance of public space design in climate change adaptation processes associated to urban flooding, an approach where efforts are targeted at assembling related disciplines and enabling their convergence. It implies a multidisciplinary practice, simply because it involves the expertise of several disciplines, such as urban planning, engineering, architecture, landscape architecture and climatology, among others. It also implies an interdisciplinary practice, since without an effective integration among the required disciplines, the adaptation measures proposed here are destined to fail their purpose. Adaptation is therefore not only challenging established professions, which previously assumed climate was generally stationary, but is now also instigating the need to redefine disciplinary competences as well as the need for intricate professional collaboration.

The need for a change of paradigm in flood risk management practices evidenced in this book is not only part of a future climate change agenda, it must also be part of the present urban agenda. Flood adaptation measures applicable in the design of public spaces promote this change of paradigm when facing urban floods. It is important for climate change adaptation advances to be achieved today by incorporating measures such as the ones here proposed alongside everyday planning practices, from municipal public space design regulations to specific requirements for urban regeneration projects.

Designing "for water" is a present matter on which the future of our cities relies.

Glossary, abbreviations, and acronyms

Glossary

Adaptation – The process of adjustment to actual or expected climate and its effects. In human systems, adaptation seeks to moderate or avoid harm or exploit beneficial opportunities. In natural systems, human intervention may facilitate adjustment to expected climate and its effects (Nouri and Matos Silva, 2013).

Albedo – Albedo is the proportion of incident solar radiation reflected by a surface. Typically given as a decimal fraction, having a value between 0 and 1 (Erell *et al.*, 2011).

Car park silo – Car park silo refers to a building dedicated for car parking. The word "silo" in this context remits to a building structure, usually cylindrical, that serves as a storage infrastructure for miscellaneous dry materials.

Catchment – The area of surface water flow contributing to a point on a drainage or river system. One catchment basin can include multiple sub-catchments.

Channel – An open conduit either naturally or artificially created that periodically or continuously contains moving water or which forms a connecting link between two bodies of water (Lehrer *et al.*, 2010).

Climate impact – The effects of climate change on natural and human systems.

Climate model – A numerical representation of the climate system that is based on the physical, chemical, and biological properties of its components and their interactions and feedback processes and that accounts for all or some of its known properties. The climate system can be represented by models of varying complexity, that is, for any one component or combination of components, a spectrum or hierarchy of models can be identified, differing in such aspects as the number of spatial dimensions, the extent to which physical, chemical, or biological processes are explicitly represented, or the level at which empirical parameterizations are involved. Climate models are applied as a research tool to study and simulate the climate, and for operational purposes, including monthly, seasonal, and inter-annual climate predictions (Intergovernmental Panel on Climate Change (IPCC), 2012).

Climate scenario – A plausible and often simplified representation of the future climate, based on an internally consistent set of climatological relationships that has been constructed for explicit use in investigating the potential consequences of anthropogenic climate change, often serving as input to impact models. Climate projections often serve as the raw material for constructing climate scenarios, but climate scenarios usually require additional information such as data about the observed current climate. (Intergovernmental Panel on Climate Change (IPCC), 2012).

Climate-driven – Related to changes in climate.

Combined sewage overflows (CSOs) – CSOs occur when, in combined sewage systems, wet stormflows exceed the sewage treatment plant capacity and are consequently directly diverted onto a receiving water body.

Combined sewage system (CSS) – Wastewater and stormwater are collected in one pipe network.

Conference of the Parties (COP) – The COP is the supreme decision-making body of the United Nations Climate Change Conference. All states that are parties to the convention are represented at the COP, at which they review the implementation of the convention and any other legal instruments that the COP adopts and take decisions necessary to promote the effective implementation of the convention, including institutional and administrative arrangements (www.unfccc.int).

Conveyance – Movement of water from one location to another.

Daylighting (streams) – Daylighting of a stream is a term that refers to the uncovering of a watercourse that was originally open-air.

First flush – First flush is the initial surface runoff of a rainstorm. During this phase, water pollution entering storm drains, and subsequently surface waters, is typically more concentrated compared to the remainder of the storm. First flush runoff typically carries a very large amount of both suspended and dissolved pollutants (Lehrer *et al.*, 2010).

Flood – The overflowing of the normal confines of a stream or other body of water, or the accumulation of water over areas not normally submerged. Floods include river (fluvial) floods, pluvial floods, coastal floods, groundwater floods, sewer floods, or artificial drainage floods, and glacial lake outburst floods.

Flood risk – In the European flood directive (2007/60/EC), flood risk is described as "the combination of the probability of a flood event and of the potential adverse consequences for human health, the environment, cultural heritage and economic activity associated with a flood event" (2007/60/EC, 2007). "Risk" is therefore the result of the function between the probability of occurrence and the consequence of the impact. While the first is dependent on climate regimes and the physical characteristics of the catchment basins by which floods are conveyed, the latter is associated to the magnitude of a flood (peak flow, volume, duration, etc.) together with the vulnerability of whatever is exposed to that particular event (Karin de Bruij *et al.*, 2009).

Greenhouse gas (GHG) – A gas in the atmosphere, of natural and human origin, that absorbs and emits thermal infrared radiation. Water vapor, carbon dioxide (CO_2), nitrous oxide (N_2O), methane (N_2O), and ozone (O_3) are the main greenhouse gases in the Earth's atmosphere. Their net impact is to trap heat within the climate system.

Groundwater – Water located beneath the ground surface in soil pore spaces and in the fractures of lithologic formations (Lehrer *et al.*, 2010).

Groundwater recharge – A hydrologic process where water moves downward from surface water to groundwater. Recharge occurs both naturally (through the water cycle) and anthropologically (i.e., "artificial groundwater recharge"), where rainwater and or reclaimed water is routed to the subsurface (Lehrer *et al.*, 2010).

Hydrological cycle – Also known as the "Water cycle", the hydrological cycle is the cycle in which water evaporates from the oceans and the land surface, is carried over the Earth in atmospheric circulation as water vapor, condenses to form clouds, and precipitates over ocean and land as rain or snow, which on land can be intercepted by trees and vegetation, provides runoff on the land surface, infiltrates into soils, recharges groundwater, discharges into streams, and ultimately flows out into the oceans, from which it will eventually evaporate again. The various systems involved in the hydrological cycle are usually referred to as hydrological systems (IPCC, 2014b).

Levee effect – The levee effect symbolizes the false sense of safety given by dams, levees, and other flood protections. These structures increase flood losses because they incite new developments in floodplains, thus promoting catastrophic damages when these man-made flood protections for some reason fail. Examples where this phenomenon was felt are numerous; among the most disastrous are: 1421, the St. Elizabeth's flood in the North Sea, the Netherlands; 1953, the North Sea floods in the Netherlands and United Kingdom; and 2005, the Hurricane Katrina floods in New Orleans.

Microclimate – A local atmospheric zone where the climate differs from the surrounding area.

Mitigation – A human intervention to reduce the sources or enhance the sinks of greenhouse gases.

Phytodepuration – Phytodepuration or phytoremediation is an ecological treatment technique that replicates natural purification processes in a controlled environment and through the use of vegetation.

POLIS Program – a Portuguese program on urban environment, specifically designed for integrating urban requalification and the improvement of the urban environment in cities.

Projection – A potential future evolution of a quantity or set of quantities, often computed by a model. Projections involve assumptions that may or may not be realized and are therefore subject to substantial uncertainty; they are not predictions.

Resilience – The capacity of social, economic, and environmental systems to cope with a hazardous event or trend or disturbance, responding or reorganizing in ways that maintain their essential function, identity, and structure. For Folke, the most resilient social-ecological systems are characterized not only by their ability to endure disturbance but also by their capacity to learn and adjust if necessary (Folke, 2006, p. 259).

Runoff – Runoff is the part of precipitation that does not evaporate and is not transpired but flows through the ground or over the ground surface and returns to the water bodies (IPCC, 2014b).

Sensitivity/Susceptibility – Sensitivity or susceptibility determines the degree to which the system is affected (beneficially or adversely wise) in regards to a given weather exposure. Sensitivity or susceptibility is typically conditioned by natural and physical conditions of the system (e.g. its topography, soils resistance to erosion, their type of occupation, among others) and the human activities that affect the natural and physical conditions of the system (e.g. agricultural practices, resource management and the forms of pressures related to settlements and population) (Dias et al. 2015).

Separate sewage system – Wastewater and stormwater are collected in two separate networks.

Singel – A particular type of Dutch canal, typically from the 19th century, that runs through the city.

Source control – The control of stormwater runoff at or near its source (Fletcher *et al.*, 2015).

Stoplogs – Hydraulic engineering control element that is used in floodgates to adjust the water level or flowrate in a river, canal, or reservoir. Stoplogs are typically long rectangular timber beams or concrete boards that are placed on top of each other and dropped into premade slots inside a weir, gate, or channel.

Storm surge – A storm surge consists in a mass of water that, through the effects of strong winds and/or the suction effects of low pressure, exceeds above the level expected from the tidal variation alone at that time and place (Intergovernmental Panel on Climate Change (IPCC), 2012).

Sustainable urban drainage systems (SUDS) or sustainable drainage systems (SuDS) – "SUDS consist of a range of technologies and techniques used to drain stormwater/ surface water. (…) They are based on the philosophy of replicating as closely as possible the natural, pre-development drainage from a site" (Fletcher *et al.*, 2015, p. 529).

Urban heat island (UHI) – The relative warmth of a city compared with surrounding rural areas. Temperatures are higher in cities than the corresponding temperatures in the surrounding rural areas. This phenomenon can be exacerbated by the large areas of low albedo surfaces, such as dark paving or roofing materials. It is generally defined as the difference between the highest air temperatures recorded in the urban canopy and the lowest recorded in the surrounding rural areas. Outlining the evolution of urban climatology since 1950, Hebbert and Jankovic highlight the fact that the urban heat island effect is the longest studied category of urban-scale climate research (Hebbert and Jankovic, 2013, p. 1343).

Vulnerability – The propensity or predisposition to be adversely affected. Vulnerability encompasses a variety of concepts and elements including sensitivity or susceptibility to harm and lack of capacity to cope and adapt (IPCC, 2014a). See concepts of sensitivity or susceptibility.

Water table – The planar, underground surface beneath which earth materials, as soil or rock, are saturated with water.

Waterbody – A body of water creating a physiographical form. A lake or a sea, for example.

Abbreviations and acronyms

ARs	Assessment reports
BMPs	Best management practices
CLABSA	Clavegueram de Barcelona, S.A.
COP	Conference of the Parties from the UNFCCC
CSO	Combined sewage overflow
CSS	Combined sewage system
ETAR	Estação de Tratamento de Águas Residuais (= WWTP)
EU	European Union
FAR	First Assessment Report
GCMs	General circulation models
GHG	Greenhouse gas
GHO	Global Health Observatory (www.who.int)
IPCC	Intergovernmental Panel on Climate Change (www.ipcc.ch)
KP	Kyoto Protocol
LID	Low impact development
LNEC	Laboratório Nacional de Engenharia Civil (National/Portuguese Civil Engineering Laboratory)
PECLAB	Plan Especial de Alcantarillado de Barcelona
RCMs	Regional climatic models
SAR	Second Assessment Report
SLR	Sea level rise
SSO	Sanitary sewage overflow
SSS	Separate sewage systems
SUDS	Sustainable urban drainage systems (SUDS) or sustainable drainage systems (SuDS)
TAR	Third Assessment Report
UHI	Urban heat island
UNFCCC	United Nations Framework Convention on Climate Change (www.unfccc.int)
WMO	World Meteorological Organization (www.wmo.int)
WSUD	Water sensitive urban design
WWTP	Wastewater treatment plant (= ETAR)

References

2007/60/EC, Directive (2007) *Directive 2007/60/EC, Assessment and Management of Flood Risks*, Brussels.

AECOM (2011) *Climate Design: Design and Planning for the Age of Climate Change: A Collection of Works from Academics and AECOM's Thought Leaders*. ORO Editions, *In Collaboration with Professor Peter Droege*. Publishers Group West, Berkeley.

AECOM (2016) *Liverpool Pier Head*. www.aecom.com/projects/liverpool-pier-head/ [accessed 14 July].

Ahern, J. (2006) Theories, methods and strategies for sustainable landscape planning. In: *From Landscape Research to Landscape Planning*. Aspects of Integration, Education and Application, Dordrecht, the Netherlands.

Ajuntament de Barcelona (2018) *Pacs i Jardins*. www.barcelona.cat/ca/que-pots-fer-a-bcn/parcs-i-jardins [accessed 31 January].

Arendt, H. (1998[1958]) *The Human Condition*. 2nd Edition. Chicago University of Chicago Press, Chicago.

Ascher, F. (2010[2001]) *Novos Principios do Urbanismo seguido de Novos Compromissos Urbanos. Um Léxico*, Margarida de Souza Lobo (trans.). 2ª edição ed. Livros Horizonte, Lisbon. Original edition, 2001. Reprint, Éditions de l'Aube.

Ashley, R.M., Faram, M.G., Chatfield, P.R., Gersonius, B. & Andoh, R.Y.G. (2010) Appropriate drainage systems for a changing climate in the water sensitive city. In: *Low Impact Development 2010: Redefining Water in the City*. ASCE, Listeria, VA, USA.

Balibrea, M.P. (2003) Memória e espaço público na Barcelona pós-industrial. *Revista Críitica de Ciências Sociais*, Dezembro, 31–54.

Banerjee, T. (2001) The future of public space: Beyond invented streets and reinvented places. *Journal of the American Planning Association*, 67(1), 9–24. doi: 10.1080/01944360108976352.

BBC (2012) *Wells-Next-the-Sea Glass Defences 'Transform' Quayside*. www.bbc.com/news/uk-england-norfolk-19208908 [accessed 22 March].

Beck, U. (1992) *Risk Society: Towards a New Modernity*. Sage Publications, Thousand Oaks, CA.

Belden, E. & Morris, K. (2011) The Elmer Avenue neighborhood demonstration project: Measuring the success of green infrastructure. *10th Annual StormCon proceedings, Anaheim, CA, US*.

Berkes, F., Colding, J. & Folke, C. (2003) *Navigating Social-Ecological Systems: Building Resilience for Complexity and Change*. Cambridge University Press, Cambridge.

Bíblia Sagrada (1991) 5th Edition. Difusora Bíblica, Lisboa.

Bicknell, J., Dodman, D. & Satterthwaite, D. (2009) *Adapting Cities to Climate Change: Understanding and Addressing the Development Challenges*. International Institute for Environment and Development (ed.). Earthscan, London.

Biesbroek, G.R. (2014) *Challenging Barriers in the Governance of Climate Change Adaptation*. Doctoral Degree, Graduate School for Socio-Economic and Natural Sciences of the Environment (SENSE), Wageningen University, the Netherlands.

Birgelen, A.von, Bornholdt, H., Brunsh, T., Böhmer, M., Funke, B., Heck, G., Heimanns, K., Mommsen, M., Klapka, A., Loidl-Reisch, C., Richter, E., Rieper, U., Rolka, C., Zadel-Sodtke, P. & Zimmermann, A. (2011) *Construir El Paisaje: Materiales, Técnicas Y Componentes Estructurales*. Zimmermann, A. (ed.). Birkhauser Architecture, Basel, Switzerland.

Boer, F., Jorritsma, J. & Peijpe, D. (2010) *De Urbanisten and the Wondrous Water Square*. 010 Publishers, Rotterdam.

Bohigas, O. (1986) *Reconstrucción de Barcelona, Monografías de la Dirección General de Arquitectura y Edificación (MOPU)*. Edicions 62, Madrid.

Borja, J. (2003) *La ciudad Conquistada*. Alianza Editorial, Madrid, Spain.

Brandão(Coord.), P., Carrelo, M. & Águas, S. (2002). *O chão da cidade. Guia de Avaliação do Design de Espaço Publico*. Lisboa: Centro Português de Design.

Brandão, P. (2004) *Ética e profissões, no Design Urbano. Convicção, responsabilidade e interdisciplinaridade. Traços da Identidade Profissional no Desenho da Cidade*. Doctoral degree, Universitat de Barcelona, Spain.

Brandão, P. (2011a) *La Imagen de la ciudad: estrategias de identidad y comunicación*. Publicacions i edicions de la Universitat de Barcelona, Barcelona, Spain.

Brandão, P. (2011b) *O Sentido da Cidade. Ensaios Sobre o Mito da Imagem Como Arquitectura, Colecção: Horizonte de Arquitectura*. Livros Horizonte, Lisboa, Portugal.

Brandão, P. (2013) Entrevista. *ArqA – Arquitectura e Arte*, mai/jun, 28–30.

Brandão, P. & Remesar, A. (2003) *Design de espaço público: deslocação e proximidade*. Centro Português de Design, Lisboa.

Brandão, P., Brandão, A., Travasso, N., Matos Silva, M., Águas, S. & Ricart, N. (2018) Public space new steet typologies: A matter of fact (Hardware and software). In: *City Street3 (CS3): Transitional Steets: Narrating Stories of Convivial Steets*. Notre Dame University, Beirut, Lebannon.

Brandi, S. (2014) Banyoles: A city between water and stone. *Estudo Prévio*, 5(arquitectoinvestigação).

Brandt, K. (2006) Hanover Kronsberg: The rainwater management concept. *Forum international Urbistique 2006 Développement urbain durable, Gestion des ressources et Services urbains, Lausanne*.

Brinke, W.ten, Karstens, S. & van Deen, J. (2010) *Sustainable Water Management under Climate Change: How to Develop Strategies for an Uncertain Future*. Deltares – Enabling Delta Life, Delft.

Broecker, W.S. (1975) Climatic change: Are we on brink of a pronounced global warming. *Science*, 189(4201), 460–463. doi: 10.1126/science.189.4201.460.

Burton, I., Challenger, B., Huq, S., Klein, R.J.T. & Yohe, G. (2001) Adaptation to climate change in the context of sustainable development and equity. In: *Climate Change 2001: Impacts, Adaptation, and Vulnerability*. Cambridge University Press, Cambridge. pp. 879–906.

Buster Simpson (2015) *Public Art Works*. www.bustersimpson.net/ [accessed 26 July].

CABE (2008) *Public Space Lessons: Adapting Public Space to Climate Change*, Taylor, D. (ed.). CABE Space, London.

Capel, H. (2005) *El Modelo Barcelona: Un Examen Crítico*. Ediciones del Serbal, Barcelona, Spain.

Carmin, J.A., Nadkarni, N. & Rhie, C. (2012) *Progress and Challenges in Urban Climate Adaptation Planning: Results of a Global Survey*. MIT Press, Cambridge, MA.

CCAP (2011) *The Value of Green Infrastructure for Urban Climate Adaptation*, Foster, J., Lowe, A. & Winkelman, S. (eds.). The Center for Clean Air Policy, Washington, DC.

ChiRoN, E. & Hidra, E. (2006) *Plano Geral de Drenagem de Lisboa. Fase A: Relatório*, Saldanha Matos, J., Silva, C., Ferreira, F., de Oliveira, R.P., Marques, R., Fonseca, T. & Branco, T. (eds.). Miraflores, Oeiras, Portugal.

ChristopherGeorge (2018) The Promenade Plantée elevated path and cycleway in Paris, France. www.shutterstock.com: Shutterstock.

City of Hanover Water Department (2000) *Water Concept Kronsberg: Part of the EXPO Project: Ecological Optimisation Kronsberg*, Altevers, B., Groß, C. & Menze, H. (eds.). Stadtentwässerung Hanover, Hanover.

CLABSA. *Clavegueram de Barcelona, S.A.* www.clabsa.es/ [accessed 4 January].

Cook, J., Nuccitelli, D., Green, S.A., Richardson, M., Winkler, B., Painting, R., Way, R., Jacobs, P. & Skuce, A. (2013) Quantifying the consensus on anthropogenic global warming in the scientific literature. *Environmental Research Letters*, 8(2), 1–7.

Corvacho, N. (2003) Câmara do Porto e Casa da Música apostadas em resolver impasse do parque de estacionamento do Castelo do Queijo. *O Público*, Local Porto. www.publico.pt/2003/02/05/jornal/camara-do-porto-e-casa-da-musica-apostadas-em-resolver-impasse-do-parque-de-estaciona-mento-do-castelo-do-queijo-197904.

Costa, J.P., Sousa, J.S., Fadigas, L., Mendes, C., Ochoa, A.R., Coelho, C.D., Fernandes, A., Serpa, F., Dias, L. & Matos Silva, M. (2013) Research project urbanized estuaries and deltas: In search for a comprehensive planning and governance: The Lisbon case. *PTDC/AUR-URB/100309/2008. Faculdade de Arquitectura – Universidade de Lisboa and Faculdade de Ciências Sociais e Humanas – Universidade Nova de Lisboa: Funded by FEDER through the Operational Competitiveness Programme – Compete and by national funds through FCT – Portuguese Foundation for Science and Technology.*

Coumou, D. & Rahmstorf, S. (2012) A decade of weather extremes. *Nature Climate Change*, 2(7), 491–496.

Couzin, J. (2008) Living in the danger zone. *Science*, 319(5864), 748–749. doi: 10.1126/science.319.5864.748.

Cowan, R. (2005) *The Dictionary of Urbanism*. Streetwise Press, Tisbury, Wiltshire.

Crutzen, P.J. (2002) Geology of mankind. *Nature*, 415(6867), 23–23.

Crutzen, P.J. & Stoermer, E. (2000) The 'anthropocene'. *Global Change Newsletter*, (41), 17–18.

Deltares. (2013) *Design Concepts of Multifunctional Flood Defence Structures*, Marlien Oderker, Dura Vermeer Business Development BV. Report Number: WP04-01-13-02. FloodProBe, The Netherlands.

Dias, L., Karadzic, V., Lourenço, T. C. & Calheiros, T. (2015) Manual para a avaliação de vulnerabilidades futuras. *Para a elaboração de Estratégias Municipais de Adaptação às Alterações Climáticas*. Edited by ClimAdaPT.Local: Portuguese Environment Agency (APA), Portuguese Carbon Fund (FPC) and EEAGrants.

De Urbanisten (2013) *Water Square Benthemplein*. www.urbanisten.nl [accessed 31 May].

Dreiseitl, H. & Grau, D. (2005) *New Waterscapes: Planning, Building and Designing with Water*. Birkhäuser Basel, Basel.

Dutch Dialogues (2011) *About Dutch 'Dialogues'*. www.dutchdialogues.com [accessed April].

EA (2011) *Temporary and Demountable Flood Protection Guide*, Ogunyoye, F., Stevens, R. & Underwood, S. (eds.). Environment Agency, Bristol, UK. Original edition, ISBN: 978-1-84911-225-3.

EEA (2012a) Climate change, impacts and vulnerability in Europe 2012: An indicator-based report. *Copenhagen, Denmark: European Environment Agency*. Original edition, ISBN 978-92-9213-346-7 doi: 10.2800/66071.

EEA (2012b) *Towards efficient use of water resources in Europe*. Copenhagen, Denmark: European Environment Agency. Original edition, doi:10.2800/95096.

Erell, E., Pearlmutter, D. & Williamson, T. (2011) *Urban Microclimate: Designing the Spaces Between Buildings*. Earthscan, London and Washington, DC.

EU Mies Award (2005) *Passeio Atlantico*. https://miesarch.com/work/2739 [accessed 30 May].

European Commission (2004) *Flood Risk Management: Flood Prevention, Protection and Mitigation*. Commission of the European Communities, Brussels, Belgium.

European Prize for Urban Public Space (2006) *Joint Winner 2006*. CCCB. www.publicspace.org/works/-/project/d078-morske-orgulje [accessed 28 May].

FashionStock.com (2012) QUEENS, NY - NOVEMBER 11: Damaged houses without power at night in the Rockaway due to impact from Hurricane Sandy in Queens, New York, U.S., on November 11, 2012. www.shutterstock.com: Shutterstock.

Fletcher, T.D., Shuster, W., Hunt, W.F., Ashley, R., Butler, D., Arthur, S., Trowsdale, S., Barraud, S., Semadeni-Davies, A., Bertrand-Krajewski, J.L., Mikkelsen, P.S., Rivard, G., Uhl, M., Dagenais, D.

& Viklander, M. (2015) SUDS, LID, BMPs, WSUD and more: The evolution and application of terminology surrounding urban drainage. *Urban Water Journal*, 12(7), 525–542. doi: 10.1080/1573062X.2014.916314.

Flood Control International (2012) *Glass Barriers: Data Sheet*. [accessed 14 May].

Flood Control International (2019) *Glass Floodwall Sea Defences: Case Studies: Wells-Next-the-Sea Glass Floodwall Installation*. www.floodcontrolinternational.com/CASE-STUDIES/case-study-wells.html [accessed 22 May].

FLOODsite (2009). Flood risk assessment and flood risk management. *An introduction and guidance based on experiences and findings of FLOODsite (an EU-funded Integrated Project)*. Deltares | Delft Hydraulics, Delft, the Netherlands.

Folke, C. (2006) Resilience: The emergence of a perspective for social-ecological systems analyses. *Global Environmental Change*, 16(3), 253–267. http://dx.doi.org/10.1016/j.gloenvcha.2006.04.002.

Forman, R.T.T. (2014) *Urban Ecology: Science of Cities*. Cambridge University Press, New York, USA.

Francisco Caldeira Cabral & Gonçalo Ribeirto Telles (1999) *A Árvore em Portugal*. 2nd Edition. Assírio & Alvim, Lisboa.

Giddens, A. (1999a) Risk and responsibility. *The Modern Law Review*, 62(1), 1–10. doi: 10.1111/1468-2230.00188.

Giddens, A. (1999b) Runaway world, lecture 2: Risk, Hong Kong. *BBC Online Network*. http://news.bbc.co.uk/hi/english/static/events/reith_99/week2/week2.htm [accessed 15 August].

Greenroofs.com (2018) *Caixa Forum Museum Vertical Garden*. www.greenroofs.com/projects/caixa-forum-museum-vertical-garden/ [accessed 7 January].

Harris, C.W., Dines, N.T. & Brown, K.D. (1998) *Time-Saver Standards for Landscape Architecture: Design and Construction Data*. McGraw-Hill Publishing Company, New York and Washington, D.C., USA.

Hartmann, T. & Driessen, P. (2013) The flood risk management plan: Towards spatial water governance. *Journal of Flood Risk Management*, 1–10. doi: 10.1111/jfr3.12077.

Hebbert, M. & Jankovic, V. (2013) Cities and climate change: The precedents and why they matter. *Urban Studies*, 50(7), 1332–1347. doi: 10.1177/0042098013480970.

Hebbert, M. & Webb, B. (2007) Towards a liveable urban climate: Lessons from Stuttgart. In: *Liveable Cities: Urbanising World: Isocarp 07*. Routledge, Manchester. pp. 1–18.

Howe, C.A. & Mitchell, C. (2012) *Water Sensitive Cities*. IWA Publishing, London, UK.

Howe, C.A., Butterworth, J., Smout, I.K., Duffy, A.M. & Vairavamoorthy, K. (2011) *Sustainable Water Management in the City of the Future*, Howe, C.A., Vairavamoorthy, K. & van der Steen, N.P. (eds.). Findings from the SWITCH Project 2006–2011 Edition. UNESCO-IHE, Delft, the Netherlands.

IBS Technics GmbH (2019) *Glass Walls*. www.flood-defenses.com/flood-protection/catastrophe-protection/glas-walls/ [accessed 22 May].

imagIN.gr photography (2009) Detail on the roof of a building in park Guell designed by Antonio Gaudi in Barcelona, Spain. www.shutterstock.com: Shutterstock.

Innerarity, D. (2006) *El nuevo espacio público*. Espasa Calpe, Madrid, Spain.

Intergovernmental Panel on Climate Change (IPCC) (2012) Managing the risks of extreme events and disasters to advance climate change adaptation. In: Field, C.B., Barros, V., Stocker, T.F., Dahe, Q., Dokken, D.J., Plattner, G.-K., Ebi, K.L., Allen, S.K., Mastrandrea, M.D., Tignor, M., Mach, K.J. & Midgley, P.M. (eds.) *A Special Report of Working Groups I and II of the Intergovernmental Panel on Climate Change*. Cambridge University Press, Cambridge, UK, and New York, NY, USA.

IPCC (Intergovernmental Panel on Climate Change) (2007) *Climate Change 2007: Synthesis Report*. Available at: http://www.ipcc.ch/pdf/assessment-report/ar4/syr/ar4_syr.pdf, Accessed 12/20/08

IPCC (2014a) *Climate Change 2014: Impacts, Adaptation and Vulnerability. Part B: Regional Aspects: Contribution of Working Groups II to the Fifth Assessment Report of the Intergovernmental Panel on Climate Change*, Barros, C.W.V.R. & Field, C.F. (eds.). IPCC, Geneva, Switzerland.

IPCC (2014b) *Climate Change 2014: Synthesis Report. Contribution of Working Groups I, II and III to the Fifth Assessment Report of the Intergovernmental Panel on Climate Change* [Core Writing Team, R.K. Pachauri and L.A. Meyer (eds.)]. IPCC, Geneva, Switzerland, 151 pp https://www.ipcc.ch/site/assets/uploads/2018/05/SYR_AR5_FINAL_full_wcover.pdf

IPCC (2014c) Summary for policymakers. In: Field, C.B., Barros, V.R., Dokken, D.J., Mach, K.J., Mastrandrea, M.D., Bilir, T.E., Chatterjee, M., Ebi, K.L., Estrada, Y.O., Genova, R.C., Girma, B., Kissel, E.S., Levy, A.N., MacCracken, S., Mastrandrea, P.R. & White, L.L. (eds.) *Climate Change 2014: Impacts, Adaptation, and Vulnerability. Part A: Global and Sectoral Aspects: Contribution of Working Group II to the Fifth Assessment Report of the Intergovernmental Panel on Climate Change*. Cambridge University Press, Cambridge, UK and New York, NY, USA.

IPCC (2015) *Organization: IPCC – Intergovernmental Panel on Climate Change*. www.ipcc.ch [accessed 16 September].

ISC (2010). Promising Practices in Adaptation & Resilience. A Resource Guide for Local Leaders. In *Climate Leadership Academy: Institute for Sustainable Communities*. Produced in partnership with Center for Clean Air Policy.

Jacobs, J. (1992[1961]) *The Death and Life of Great American Cities*. Random House, New York.

Kew, B., Pennypacker, E. & Echols, S. (2014) Can greenwalls contribute to stormwater management? A study of cistern storage greenwall first flush capture. *Journal of Green Building*, 9(3), 85–99. doi: 10.3992/1943-4618-9.3.85.

Klimakvarter (2019) *Taasinge Plads*. http://klimakvarter.dk/ [accessed 11 June].

Knippers Helbig Advanced Engineering (2010) Megastructure expo Shanghai 2010. *Stahlbau*, 79(5), 400–402. doi: 10.1002/stab.201090040.

Kravčík, M., Pokorný, J., Kohutiar, J., Kováč, I.M. & Tóth, E. (2007) *Water for the Recovery of the Climate: A New Water Paradigm*. Publication printed with the financial support of: Municipalia a. s. and TORY Consulting a.s. Edition. Krupa Print, Žilina.

Kwadijk, J.C.J., Haasnoot, M., Mulder, J.P.M., Hoogvliet, M.M.C., Jeuken, A.B.M., van der Krogt, R.A.A., van Oostrom, N.G.C., Schelfhout, H.A., van Velzen, E.H., van Waveren, H. & de Wit, M. J.M. (2010) Using adaptation tipping points to prepare for climate change and sea level rise: A case study in the Netherlands. *Wiley Interdisciplinary Reviews: Climate Change*, 1(5), 729–740. doi: 10.1002/wcc.64.

Kwon, K.-W. (2007) Cheong Gye Cheon Restoration Project, a revolution in Seoul. *Seoul Metropolitan Government*. http://worldcongress2006.iclei.org [accessed 2 April].

Landscape Institute (2014) Case studies library. *Inspiring Great Places*. www.landscapeinstitute.org/casestudies/ [accessed 22 Mach].

Lehrer, M., Margulies, E., Dyer, J., Link, J., Goetting, D., Birkeland, B., Hall, J., Leventhal, R. & Stromberg, E. (2010) *Compton Creek*. Earthen Bottom Enhancement Feasibility Study. Coastal Conservancy, City of Compton.

Lennon, M., Scott, M. & O'Neill, E. (2014) Urban design and adapting to flood risk: The role of green infrastructure. *Journal of Urban Design*, 19(5), 745–758. doi: 10.1080/13574809.2014.944113.

Lister, N.-M. (2005) Ecological design for industrial ecology: Opportunities for re(dis)covery. In: Coté, R., Tansey, J. & Dale, A. (eds.) *Linking Industry and Ecology: A Question of Design*. UBC Press. Vancouver, Toronto, pp. 15–28.

LLDC (2016). *Landscaping the Park*. http://queenelizabetholympicpark.co.uk/. [accessed 22 January]

LNEC (1983) *Contribuição para o estudo da drenagem de águas pluviais em zonas urbanas*. Volume 2, Abreu, M.R. (ed.). Laboratório Nacional de Engenharia Civil, Lisboa, Portugal.

Lynch, K. (1996[1960]) *A imagem da cidade*. edições 70, Lisboa.

Madanipour, A. (1997[1972]) Ambiguities of urban design. *Town Planning Review*, 68(3), 363–383.

Mailliet, L. & Bourgery, C. (1993) *L'arboriculture urbaine*, Institut pour le développement forestier (ed.). Collection Mission du paysage, Paris, France.

Margolis, L. & Robinson, A. (2007) *Living Systems: Innovative Materials and Technologies for Landscape Architecture*. Birkhäuser, Basel, Boston, Berlin.

Marshall, R. (2001) *Waterfronts in Post-Industrial Cities*, Marshall, R. (ed.). Spon Press, New York, USA.

Martin, L. (2007) The grid as generator. In: Carmona, M. & Tiesdell, S. (eds.) *Urban Design Reader*. Architectural Press, Oxford, UK. Original edition, 1972. pp. 70–82.

Matos Silva, M. (2011) *El Modelo Barcelona de espacio público y diseño urbano: Public Space and Flood Management/Dipòsits d'aigües pluvials*. Master Degree, Master Oficial en Disseny Urbà: Art, Ciutat, Societat, Universitat de Barcelona, Spain.

Matos Silva, M. (2016) *Public Space Design for Flooding: Facing the Challenges Presented by Climate Change Adaptation*. Doctoral degree, Departament d'Escultura, Universitat de Barcelona, Spain.

Matos Silva, M. & Costa, J.P. (2016) Flood adaptation measures applicable in the design of urban public spaces: Proposal for a conceptual framework. *Water*, 8(7), 284–310.

Matos Silva, M. & Costa, J.P. (2017) Urban flood adaptation through public space retrofits: The case of Lisbon (Portugal). *Sustainability*, 9(5), 816.

Matos Silva, M. & Costa, J.P. (2018) Urban floods and climate change adaptation: The potential of public space design when accommodating natural processes. *Water*, 10(2), 180.

Matos Silva, M. & Nouri, A. (2014) Adaptation measures on riverfronts, an overview of the concepts. In: Costa, J.P. & de Sousa, J.F. (eds.) *Climate Change Adaptation in Urbanised Estuaries Contributes to the Lisbon Case*. Faculdade de Ciências Sociais e Humanas da Universidade Nova de Lisboa, Óbidos. pp. 131–150.

Matos Silva, M. & Rego, J.S. (2019) Sistemas ecológicos na grande escala: o caso do Eixo Verde e Azul. In: Brandão, A. & Brandão, P. (eds.) *O lugar de todos. Interpretar o espaço público urbano*. IST-ID, Associação do Instituto Superior Técnico para a Investigação e Desenvolvimento, Lisboa. pp. 39–41.

Mias Arquitectes (2008) *Banyoles Old Town Public Space*. www.miasarquitectes.com/portfolio/banyoles-old-town/ [accessed 14 June].

Mital, S. (2013) A rain catcher whose mission is to simplify RWH. *The New Indian Express*, 1–2.

mvvainc (2012) Projects: Corktown common. *Michael van Valkenburgh Associates*. www.mvvainc.com/ [accessed 18 July].

Neue Ufer (2013) *Home*. www.neue-ufer.de [accessed 16 July].

Nouri, A.S. & Matos Silva, M. (2013) Climate change adaptation and strategies: An overview. In: Bártolo, H., Da Silva Bartolo, P.J., Alves, N.M.F., Mateus, A.J., Almeida, H.A., Lemos, A.C. S., Craveiro, F., Ramos, C., Reis, I., Durão, L., Ferreira, T., Duarte, J.P., Roseta, F., Costa, E.C. E., Quaresma, F. & Neves, J.P. (eds.) *Green Design, Materials and Manufacturing Processes*. Taylor and Francis, Lisbon. pp. 501–507.

Novotny, V., Ahern, J. & Brown, P. (2010) *Water Centric Sustainable Communities: Planning, Retrofitting, an Building the Next Urban Environment*. John Wiley & Sons, Inc, Hoboken, NJ.

Oke T.R. (1997) Part 4: the changing climatic environments: Urban climates and global environmental change. In: Thompson, R.D. & Perry, A. (eds.), *Applied Climatology Principals and Practice*. Routledge, London. pp. 273–287.

Oorschot, L. (2013) Regeneration of the urban coastal area of Scheveningen: Pearl by the sea. *TRIA – Territorio della Ricerca su Insediamenti e Ambiente. Rivista internazionale di cultura urbanistica*, 11(2), 171–184. https://doi.org/10.6092/2281-4574/2045.

Oppenheimer, M. (2010) Ice sheets, sea level rise, and the increasing risk to deltas. *Key Speech at 'Deltas in Times of Climate Change' International Conference*. www.climatedeltaconference.org/results [accessed 25 January].

Papacharalambous, M., Davis, M.S., Marshall, W., Weems, P. & Rothenberg, R. (2013) *Greater New Orleans Urban Water Plan: Implementation*. Waggonner & Ball Architects, New Orleans, LA, USA.

Pardal, S. (2006) *Parque da Cidade do Porto – Ideia e Paisagem*. 2nd Edition. GAPTEC – Gabinete de Apoio da Universidade Técnica de Lisboa, Lisboa, Portugal.

Pelling, M. (1997) What determines vulnerability to floods: A case study in Georgetown, Guyana. *Environment and Urbanization*, 9(1), 203–226. doi: 10.1177/095624789700900116.

Philip, R. (2011) *Switch Training Kit: Integrated Urban Water Management in the City of the Future*, Philip, B.A.R. & Loftus, A.-C. (eds.). Volume Module 4 Stormw ater: Exploring the options. ICLEI European Secretariat GmbH | Gino Van Begin, Freiburg, Germany.

Pinho, P., Martins, A., Costa, J.P., Dias, L., Cruz, S.S., Sousa, S., Morgado, S. & Oliveira, V. (2008) *Supercities: Sustainable Land Use Policies for Resilient Cities*, Pinho, P. (ed.). R&D Project financed by the Portuguese Foundation for Science and Technology (ref.: URBAN/AUR/0002/2008); University of Porto and University of Lisbon, Porto and Lisbon, Portugal.

Pinto, A.J. (2015) *Coesão urbana: o papel das redes de espaço público*. Doctoral degree, Departament d'Escultura, Universitat de Barcelona, Spain.

Pinto, A.J., Remesar, A. & Brandão, P. (2011) Networks and anchors: From morphology to the strategy of urban cohesion. *Urban Morphology in Portugal: Approaches and Perspectives. 1st Portuguese Network on Urban Morphology, Lisbon*.

Pope Francis (2015) Encyclical letter Laudato Si' of the Holy Father Francis. *Vatican Press*. http://w2.vatican.va/content/vatican/en.html [accessed 5 October].

Portas, N. (2003) Espaço Público e a cidade emergente – Os novos desafios. In: Brandão, P. & Remesar, A. (eds.) *Design de espaço público: deslocação e proximidade*. Centro Português de Design, Lisboa, Portugal. pp. 16–19.

Portas, N. (2011[1968]) *A Cidade Como Arquitectura*. Livros Horizonte, Lisbon.

Portas, N. (2012) O Urbano e a Urbanística ou os tempos das formas. [Video]. *Culturgest*, Last Modified 28 of January. www.culturgest.pt/actual/01/01-nunoportas.html, https://vimeo.com/56981101 [accessed 29 January 2012].

Pötz, H. & Bleuzé, P. (2012) *Urban green-blue grids for sustainable and dynamic cities*. Coop for Life, Delft.

PPS (2003) What makes a successful place?. *Project for Public Spaces*. www.pps.org/reference/grplacefeat/ [accessed 27 June].

Programa Polis Cacém (2000) *Viver o Cacém – Plano Estratégico*, Programa de Requalificação Urbana e Valorização Ambiental de Cidades (eds.). Ministério do Ambiente e Ordenamento do Território, Lisboa.

Prominski, M., Stokman, A., Zeller, S., Stimberg, D. & Voermanek (2012) *River Space Design: Planning Strategies, Methods and Projects for Urban Rivers*, Stein, R. (ed.) Birkhauser, Basel, Switzerland.

Queen Elizabeth Olympic Park (2018) Documents. *London Legacy Development Corporation*. www.queenelizabetholympicpark.co.uk/our-story/publications/documents [accessed 31 May].

RCI (2009). *Rotterdam Climate Proof Adaptation Programme*. The Rotterdam Challenge on Water and Climate Adaptation. Rotterdam Climate Initiative. City of Rotterdam.

Read, C. 2012[1898]. *Logic, Deductive and Inductive*. London: Tradition Classics.

Read, C. (2012[1898]) *Logic, Deductive and Inductive*. Tradition Classics, London.

Remesar, A. (2005) Reflexiones sobre la privatización del espacio público. *intervir*, 32–37.

Ribeiro, L. & Barão, T. (2006) Greenways for recreation and maintenance of landscape quality: Five case studies in Portugal. *Landscape and Urban Planning*, 76(1–4), 79–97. http://dx.doi.org/10.1016/j.landurbplan.2004.09.042.

Ricart, N. & Remesar, A. (2013) Reflexiones sobre el Espacio Publico. *on the w@terfront*, 25, 5–35.

Robinson, A. & Hopton, H.M. (2011) Case study of elmer avenue neighborhood retrofit. *Landscape Architecture Foundation*. http://landscapeperformance.org/ [accessed 12 January].

Ruddell, D., Harlan, S., Grossman-Clarke, S. & Chowell, G. (2012) Scales of perception: Public awareness of regional and neighborhood climates. *Climatic Change*, 111(3–4), 581–607. doi: 10.1007/s10584-011-0165-y.

Scholz, M. & Grabowlecki, P. (2007) Review of permeable pavement systems. *Building and Environment*, 42(11), 3830–3836. doi: 10.1016/j.buildenv.2006.11.016.

Seelemann, G. (2015) Opening-up the rivers in Leipzig. *On Behalf of Development Association Neue Ufer Leipzig e.V.* www.urban-project.lviv.ua/ [accessed 16 July].

Sliedrecht, M., Molenaar, A., Jacobs, J., van der Vlies, A., Helmer, J., Slooters, G., van de Esschert, M., Vos, B. & Kuijpers, J. (2007) *Waterplan 2 Rotterdam: Working on Water for an Attractive City*, Jacobs, J., de Greef, P., Bosscher, C., Haasnoot, B., Wever, E., Speelman, J.P. & de Jong, M. (eds.). Municipality of Rotterdam, Schieland and Krimpenerwaard Water Control Board, Hollandse Delta Water Authority, Delfland Water Control Board, Rotterdam.

Smith, M. (2010) Moving edges. *Key Speech at 'Deltas in Times of Climate Change' International Conference*. www.climatedeltaconference.org/results [accessed 25 January].

Solà-Morales, M.de (2012) *Scheveningen Den Haag, 2006–2012*. http://manueldesola-morales.com [accessed 26 January].

Stamać, I. (2005) Acoustical and musical solution to wave: Driven sea organ in zadar. *2nd Congress of Alps-Adria Acoustics Association and 1st Congress of Acoustical Society of Croatia, 23–24 June, Opatija, Croatia.*

Stedman R.C. (2004) Risk and climate change: Perceptions of key policy actors in Canada. *Risk Analysis*, 24,1395–1406.

Steers, J.A., Stoddart, D.R., Bayliss-Smith, T.P., Spencer, T. & Durbidge, P.M. (1979) The storm surge of 11 January 1978 on the East Coast of England. *The Geographical Journal*, 145(2), 192–205. doi: 10.2307/634386.

Stefulesco, C. (1993) *L'urbanisme vegetal, Collection Mission du Paysage*. Institut pour le développement Forestier, Paris, France.

Stimson Studio (2017) *UMass Design Building: Amherst, MA*. www.stimsonstudio.com/umass-design-building [accessed 28 May].

Susdrain (2019) Susdrain case studies. *CIRIA*. www.susdrain.org/case-studies [accessed 31 May].

Te Linde, A. & Jeuken, A. (2011) *Werken met knikpunten en adaptatiepaden, handreiking*. Deltares report nr. 1202029-000.

Trip, J.J. (2007) *What Makes a City? Planning for 'Quality of Place': The Case of High-Speed Train Station Area Redevelopment*, Delft Cetre for Sustainable Urban Areas (ed.), *Sustainable Urban Areas*. IOS Press, Delft University Press, the Netherlands.

Urban, J. (2008) *Up By Roots*. International Society of Arboriculture, Champaign, IL.

Valera, S. (2001) La percepció del risc. In: Mir, N. (ed.) *Com 'sentim' el risc*. Beta Editorial, Barcelona. pp. 235–261.

Van Der Linden, S. (2014) Towards a new model for communicating climate change. In: Cohen, S., Higham, J., Peeters, P. & Gössling, S. (eds.) *Understanding and Governing Sustainable Tourism Mobility: Psychological and Behavioural Approaches*. Routledge, Taylor and Francis Group, Abingdon, UK. pp. 243–275.

Veelen, P.C.van (2013) *Adaptive Strategies for the Rotterdam Unembanked Area, Synthesis Report*. National Research Programme Knowledge for Climate, the Netherlands.

Vidiella, À.S. & Zamora, F. (2011) *Paisajismo Urbano – Barcelona – Urban Landscape*, de Barcelona, A. (ed.). Imatge i Serveis Editorials Municipals, Barcelona.

Voorendt, M.Z. (2015) Examples of multifunctional flood defences: Working report. *TU Delft University of Technology: Department of Hydraulic Engineering*, Delft, The Netherlands.

WATERFRONToronto (2009) *Explore Projects: Corktown Common"* www.waterfrontoronto.ca [accessed 18 July].

White, I. & Howe, J. (2004) The mismanagement of surface water. *Applied Geography*, 24(4), 261–280. doi: 10.1016/j.apgeog.2004.07.004.

Wilbanks, T.J. & Kates, R.W. (1999) Global change in local places: How scale matters. *Climatic Change*, 43(3), 601–628. doi: 10.1023/A:1005418924748.

Zaitzevsky, C. (2001) The 'Emerald Necklace': An historic perspective. In: Walmsley, T. & Pressley, M. (eds.) *Emerald Necklace Parks: Master Plan*. Commonwealth of Massachusetts, Department of Environmental Management, Boston, MA, USA. pp. 27–42.

Zheng, X. (2007) The ray and stata center. *TOPOS The International Review of Landscape Architecture and Urban Design, n°59. Water: Design and Management*, 45–49, Munich, Germany.

Index

Note: Page numbers in **bold** indicate a table on the corresponding page.

Sustainable Cities Research Series